Luis de Sousa

Plans, Elevations, Sections and Views of the Church of Batalha in the Province of Estremadura in Portugal

Luis de Sousa

Plans, Elevations, Sections and Views of the Church of Batalha in the Province of Estremadura in Portugal

ISBN/EAN: 9783337764074

Printed in Europe, USA, Canada, Australia, Japan

Cover: Foto ©berggeist007 / pixelio.de

More available books at **www.hansebooks.com**

London, May 31, 1792.

TO THE ENCOURAGERS OF THE POLITE ARTS.

THIS DAY IS PUBLISHED, No. I. OF A NEW WORK,
IN THE

GOTHIC STYLE OF ARCHITECTURE,
DEDICATED TO
THE RIGHT HONOURABLE WILLIAM CONYNGHAM.

PROPOSALS
FOR PUBLISHING BY SUBSCRIPTION,

DESIGNS OF THE CHURCH AND ROYAL MONASTERY OF
BATALHA,
Including the MAUSOLEUM of King JOHN I. and King EMANUEL,
Situate in the Province of ESTRE-MADURA, in PORTUGAL;

Measured and Drawn on the Spot, in the Year 1789,

BY JAMES MURPHY, ARCHITECT.

WITH

An HISTORICAL and DESCRIPTIVE ACCOUNT of this famous Gothic Structure;
Translated from the Portuguese of FR. LUIS DE SOUSA.
With REMARKS and OBSERVATIONS by the AUTHOR.

TERMS OF THE SUBSCRIPTION.

THE Work will consist of Five Numbers in Imperial Folio, each containing from Four to Five Plates, beside the Letter-Press.

THE Price to Subscribers will be Half-a-Guinea each Number, or Fifteen Shillings elegantly Printed on superfine Vellum Paper, beside Half-a-Guinea to be paid at the Time of Subscribing.

EACH Number to be paid for on Delivery, which will take place in the Order the Work is subscribed for.

FROM the unexpected Delays that occurred in the Engraving of the Plates of this Work, the Author thought it advisable to publish it in smaller Numbers than what he first proposed, which was to have had Eight Plates in each Number, at One Guinea each; here the Number of Plates are from Four to Five, at half that Price; to which Plan he hopes his Subscribers can have no Objection.—The Remainder of the Work is in great Forwardness, and will be delivered with all convenient Expedition.

A Title Page, Etched by the Author, will be given with one of the subsequent Numbers, together with the Preface and Directions for arranging the Plates according to their respective Descriptions.

ANY of the Plates of the Work may be had of the Author, at the Price affixed to each.

SUBSCRIPTIONS ARE RECEIVED BY

Mr. CADELL, Strand,	Messrs. TAYLORS, No. 56, High Holborn,
Mr. EDWARDS, Pall Mall,	Mr. MACKLIN, Fleet-Street.
AND BY THE AUTHOR.	

PLANS ELEVATIONS SECTIONS and VIEWS of the CHURCH of BATALHA, in the PROVINCE of ESTREMADURA in PORTUGAL, with the History and Description by FR. LUIS DE SOUSA, with remarks. To which is prefixed an INTRODUCTORY DISCOURSE on the PRINCIPLES of GOTHIC ARCHITECTURE by JAMES MURPHY.

LONDON Printed for I. & J. TAYLOR, High Holborn MDCCXCV.

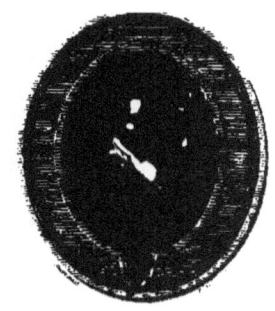

TO
THE RIGHT HONOURABLE
WILLIAM CONYNGHAM
ONE OF HIS MAJESTY'S MOST HONOURABLE PRIVY COUNCIL
TELLER OF THE EXCHEQUER IN IRELAND
TREASURER OF THE ROYAL IRISH ACADEMY
FELLOW OF THE SOCIETY OF ANTIQUARIES LONDON, &c. &c.
THIS WORK IS HUMBLY DEDICATED

BY HIS MOST OBEDIENT
AND VERY MUCH OBLIGED
HUMBLE SERVANT
LONDON MAY III MDCCXCII.   JAMES MURPHY.

PREFACE.

THE Royal Monastery of Batalha, the subject of the following Work, is a structure very little known, though the excellence of its architecture justly entitles it to rank with the most celebrated Gothic edifices of Europe. My first knowledge of this venerable pile, was derived from seeing some sketches of it in the possession of the Right Honourable William Conyngham, taken by himself, and two other gentlemen * who travelled with him through Portugal in the year 1783. These sketches, which are very correct representations of the original, gave me so high an idea of that building, as to excite in me an earnest desire to visit it; and the above Gentleman having generously offered me his patronage and support, I set out from Dublin, in a trading vessel, and arrived at Oporto in the month of January 1789. Whence I departed after a short stay, and in seven days reached Batalha, where I was kindly received by the Prior and all the Convent.

This Monastery is situated in a small village of the same name, in the province of Estremadura in Portugal, about sixty miles N. of Lisbon; it is environed by mountains, some of which produce pine and olive trees. The country around it is pretty well cultivated, particularly the plains, which are naturally fertile, and well watered, yielding great quantities of excellent grapes and olives. The village is inhabited chiefly by indigent but industrious people, many of whom derive a comfortable subsistence from their employments in the service of the Convent.

The building, considering its age, is in good preservation, and has suffered very little from the usual injuries of the elements; owing to the durability of the materials, and the serenity of the climate. Some parts, however, have been damaged by the fatal earthquake of 1745, and the spire that crowned the Mausoleum of the founder (King John I.) was entirely destroyed by that disaster, but fortunately, in its fall, did not hurt any part of the inside. This spire has not been rebuilt; the other parts of the Mausoleum which received most injury, have been decently repaired in their former style, through the munificence of His Most Faithful Majesty the late King, Joseph I. Some of the pinnacles, and parts of the railing of the Church, were also thrown down, or deranged, at the same time; and had not been replaced when these observations were made, in 1789. This however we must attribute, not to any neglect on the part of the Fathers, but to the poor revenue of the Convent; for, to do them justice, they hold this edifice in great veneration, and, as far as circumstances will allow, are careful in repairing and cleaning it. They cannot, indeed, completely repair past damages, but they use every precaution in their power against future ones.

* Colonel Tarrant, and Captain Broughton.

In the Church belonging to this Monastery, we observe none of those trifling and superfluous sculptures, which but too often are seen to crowd other Gothic edifices. Whatever ornaments are employed in it, are sparingly, but judiciously disposed; particularly in the inside, which is remarkable for a chaste and noble plainness; and the general effect, which is grand and sublime, is derived, not from any meretricious embellishments, but from the intrinsic merit of the design. The forms of its mouldings and ornaments, are also different from those of any other Gothic building that I have seen. This difference chiefly consists in their being turned very quick, and cut sharp and deep; with some other peculiarities, which the plates of this work will sufficiently explain. Throughout the whole are seen a correctness and regularity, which evidently appear to be the result of a well conceived original design; it is equally evident that this design has been immutably adhered to, and executed in regular progression, without those alterations and interruptions to which such large buildings are commonly subject.

These and other considerations equally interesting, induced me to measure and delineate the whole, with as much accuracy as possible; which I completed in thirteen weeks; during which time I lodged and boarded in the Convent. I am happy in this opportunity of returning my most sincere thanks to the Prior, the *Hospedeiro Mor*, and the rest of the Fathers of Batalha, for the politeness and attention they constantly shewed me. The piety, hospitality, and simplicity of these reverend Fathers, can scarcely be imagined in these degenerate times; they call to our recollection the description historians give us, of the Christians of the Apostolic ages; their sanctity of manners increases the dignity of the venerable mansion they inhabit.

I cannot conclude without acknowledging my obligations to the Right Honourable William Conyngham, by whose munificence I have been enabled to carry on this work. The Portuguese have too much gratitude, not to add their acknowledgments to him also, for having made known the merits of this inimitable structure. Till now, no part of it, so far as I could learn, has ever been published. The honour of presenting it to the world, was reserved for a private Gentleman, a native of Ireland, who, induced by no other motive than a love of the fine Arts, and a wish for the advancement of Science, has expended upwards of One Thousand Pounds, in rescuing this noble edifice from the obscurity in which it has lain concealed for ages.—I have taken the liberty to dedicate this Work to him, in consideration of his exemplary liberality, and as an humble testimony of my everlasting gratitude and respect.

ARCHES · OF · VARIOUS · KINDS:

INTRODUCTION PLATE II.

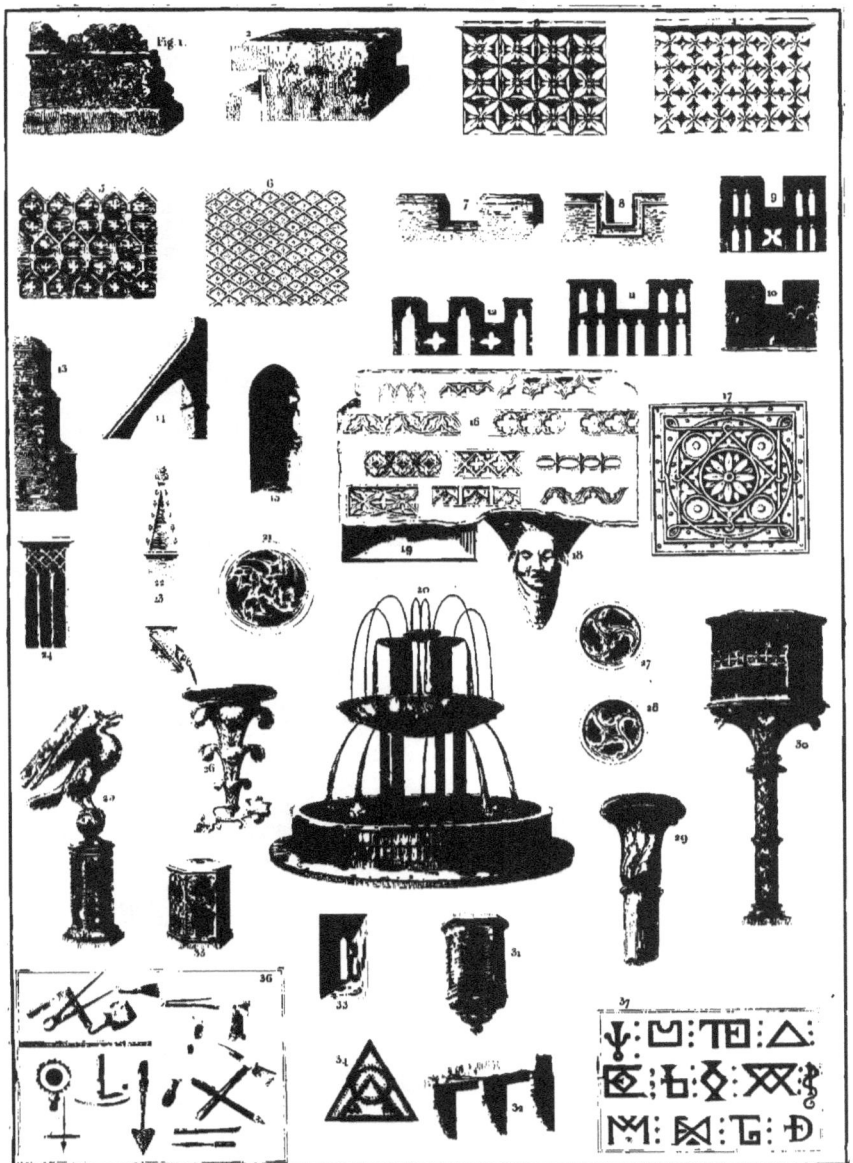

FRAGMENTS OF GOTHIC ARCHITECTURE.

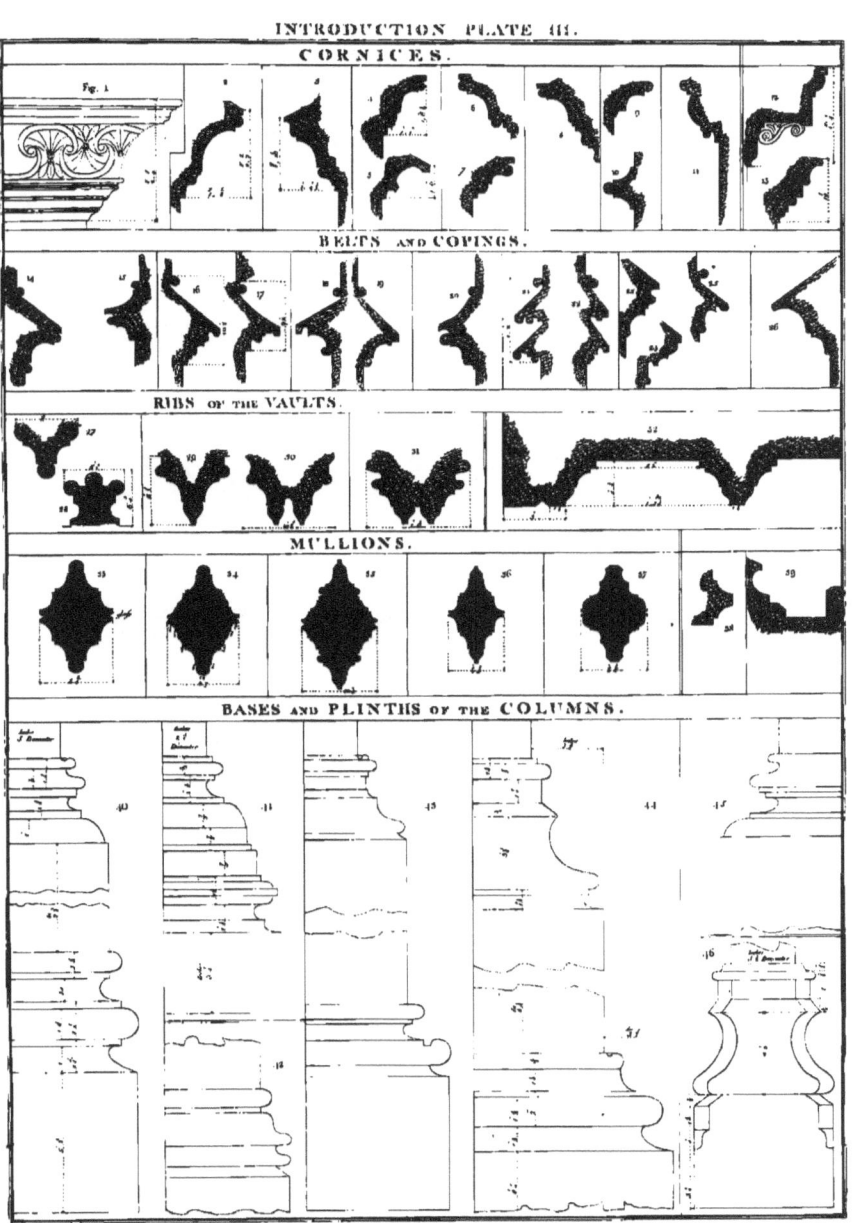

FRAGMENTS OF GOTHIC ARCHITECTURE.—BATALHA.

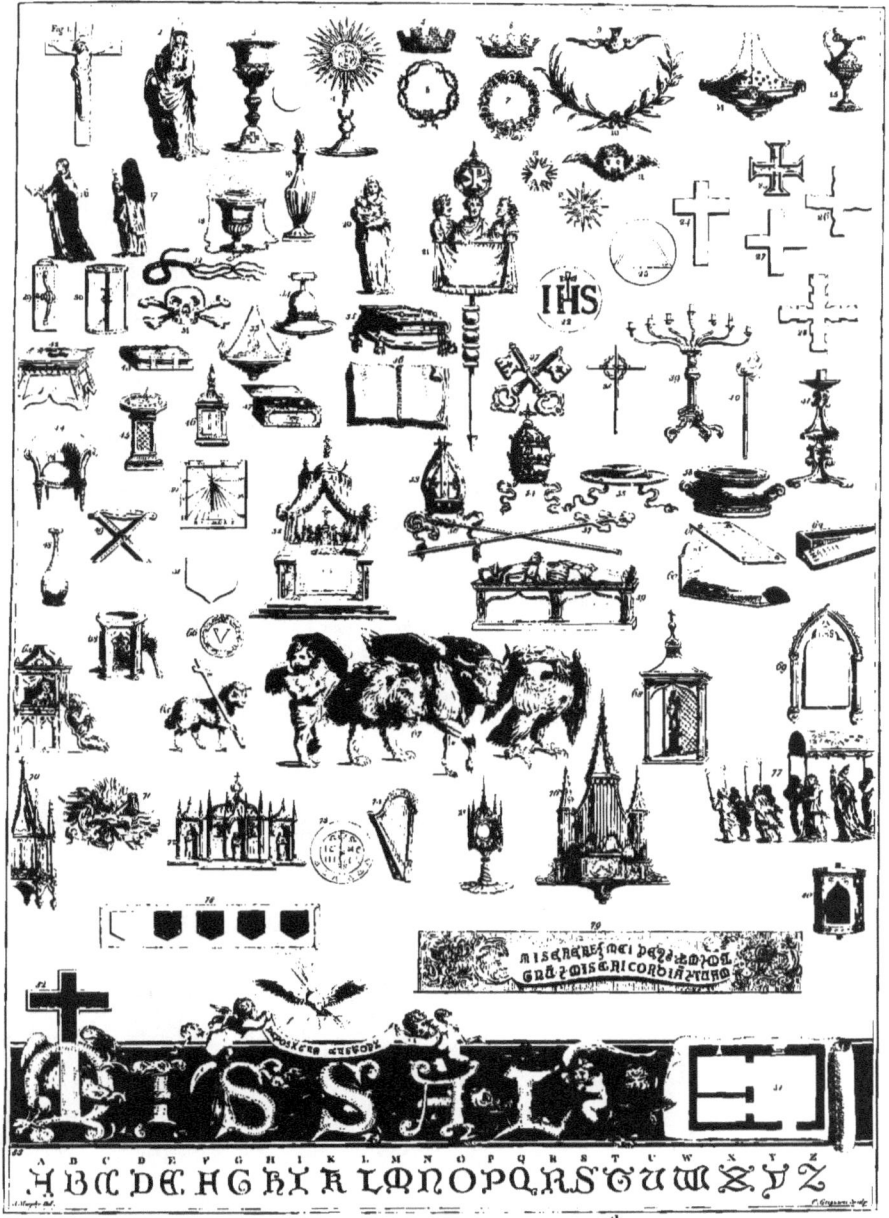

RELIGIOUS CUSTOMS OF THE 13th CENTURY.

INTRODUCTION.

WHILST the remains of the edifices of ancient Greece and Rome, have been measured and delineated with the greatest accuracy, by many persons well qualified for the task, very few have directed their enquiry towards the principles of that style of architecture called Gothic. This neglect may, in a great measure, be attributed to a prejudice arising from a mistaken notion of its having originated with a tribe of barbarians, from whom nothing excellent could be expected; but there is no reason to suppose, that they have any claim to the invention of that elegant species of it which is exhibited in the following work. This species is allowed by the most competent judges, to have originated with the Normans, towards the conclusion of the twelfth century, and is generally known by the name of the Modern Norman Gothic.

Many ingenious men, who have carefully examined the best churches executed in this style, allow they are highly deserving of attention, whether we consider them as vestiges of art, or as monuments of the industry and manners of former ages. An eminent Artist of the present time, who is very competent to judge of their merit, remarks that, " to those usually called Gothic Architects, we are indebted for the first considerable improvements in construction; there is a lightness in their works, an art and boldness of execution, to which the ancients never arrived, and which the moderns comprehend and imitate with difficulty." *

These circumstances might be considered as sufficient inducement to investigate the principles on which those edifices are built, but they have other claims to our notice. No other mode of building is so well calculated to excite sublime and awful sensations; and if we admire the heathen temples of ancient Greece and Rome, because they awaken these emotions in us, we must surely esteem these Christian temples, as they certainly produce that effect in a superior degree, though, in the construction of most of them, nothing is used but the mere produce of the quarry. Of this we have a striking instance in the church of Batalha, which is formed out of as simple materials as the great pyramid of Egypt; yet the simplicity of its matter, but augments the gracefulness of its appearance.

* Sir William Chambers on Civil Architecture, page 14, third edition.

INTRODUCTION.

When we consider the difficulty the Architects of these edifices had to contend with, from the ignorance of the times, and the debased state of every Art and Science, we must confess they had more merit than is generally allowed them; for, notwithstanding these powerful barriers, their works discover signs of mathematical knowledge, of philosophical penetration, and of religious sentiments, which future generations may perhaps seek for in vain, in the productions of the Architects of this enlightened age.

The earliest specimens of this manner of building in England were, I believe, finished about the beginning of the thirteenth century; and though, perhaps, not constructed in imitation of any ancient models, were carried to a greater pitch of excellence in less time, than history records of any other species of Architecture; and may be mentioned as remarkable instances of the vigorous exertions of the human mind, in the early stages of an art. It must however be expected, that some defects are to be found in these edifices, as perfection, in any art, is a plant of slow growth. But if this mode of building had still continued to be cultivated, with that ardour which marked its progress in this country, a little before the period of its final dissolution; improved by the assemblage of various scattered perfections, and graced by emanations from the kindred arts, what excellence might we not reasonably expect to have seen it brought to, when the excrescences, and inelegancies, of ruder times, would have been lopped off by the hand of taste.

The following work exhibits one of the best specimens existing of the Gothic style, in which I have attempted to follow the manner practised by former artists, with so much success, in describing and delineating the ancient edifices of Rome; and though it cannot be expected that this single specimen, however excellent, will be sufficient to ascertain completely the principles on which Gothic Edifices are built, yet, I trust that it will help to develop those intricacies of construction, which no one has hitherto attempted to explain, and that with the assistance of other attempts of the same nature, we may be led to comprehend the mysterious rules of the Reverend Fathers who are supposed to have been their Architects.*

The writers who have hitherto treated on this subject, have principally directed their attention to the POINTED ARCH, which they seem to consider as the leading

---

* Bentham's Ely, p. 27.————Monsieur Felibien, speaking of the Architecture of the sixth Century, says, "in constructing the different edifices I have mentioned, they employed no workmen scarcely but masons, who had no further knowledge of the sciences than the practice of preparing the mortar well, and of choosing good materials; in which, indeed, they took such precautions, that nothing can be more solid than their works. I do not, however, rank such persons among the number of Architects. I believe that there were very few lay-men deserving this title under our first kings, as most of them were at that time occupied in what related to the profession of war, leaving to the churchmen the care of cultivating the Sciences, and the fine Arts. What strengthens this opinion respecting Architecture, is, that in France the first monks worked themselves in building their monasteries, employing the most intelligent of their community to conduct these works, without the assistance of the laity; nay, even the superiors were often at the head of their monks, to give the designs, and to trace the form in which the stones were to be cut. This employment was so far from degrading the ecclesiastic dignity, that many of the bishops thought it an honour to be reputed the architects, and overseers of the churches which they built, imitating in that respect, the High-priests of the Jewish law, who, it is said, employed themselves in building and repairing the Temple of Jerusalem."

Entretiens sur la vie des Architectes, tom. v. p. 167.

INTRODUCTION.

characteristic of this species of Architecture. Many disquisitions have been written concerning its origin, but it still remains unexplained. I have bestowed much thought on this part, and flatter myself, that though the conjectures I am now about to offer respecting its origin are entirely new, they will upon mature consideration, be allowed to approach as near to certainty as the nature of the subject will admit.

If the Pointed Arch be considered detached from the building, its origin may long be sought for in vain, and indeed I imagine that this is the reason it has eluded the researches of so many ingenious men; but, on the contrary, if we examine it in a relative view, as a part in the composition of the whole, it will become more easy to account for its form, or for that of any other component part. If we take a comprehensive view of any of these structures externally, we shall perceive that not only the arch, but every vertical part of the whole superstructure terminates in a point. And the general form, if viewed from any of the principal entrances, (the station from whence the character of an edifice should be taken) will be found to have a pyramidal tendency. The Porticos of the first story, whether they be three, or five in number, are reduced to one at the top, and this is sometimes crowned with a lofty pediment, which might more properly be called a pyramid, as we see in the transept front of Westminster Abbey, and York Minster. If we look further on, in a direct line with its apex, we frequently see a lofty spire, or pyramid, rising over the intersection of the nave and transept. Each of the buttresses and turrets is crowned with a small pyramid. If niches are introduced they are crowned with a sort of pyramidal canopy. The arches of the doors and windows terminate in a point; and every little accessory ornament, which enriches the whole, has a pointed or angular tendency. Spires, pinnacles, and pointed arches;* are always found to accompany each other, and very clearly imply a system founded on the principles of the Pyramid.

It appears evident, from these instances, that the pyramidal form actually exists throughout the several component parts, and the general disposition of the edifice, approaches as near to it, at least, as the ordonnance of an historical painting which is said to be pyramidally grouped. Hence we may comprehend the reason why the arch was made pointed, as no other form could have been introduced with equal propriety, in a pyramidal figure, to answer the different purposes of uniformity; fitness; and strength. It is in vain, therefore, that we seek its origin in the branches of trees; or in the intersection of *Saxon* or *Grecian circles* ; or in the perspective of arches ; or in any other accidental concurrence of fortuitous

---

* "As for spires and pinnacles with which our oldest churches are sometimes, and more modern ones are frequently decorated, I think they are not very ancient. The towers and turrets of churches, built by the *Normans* in the first century after their coming, were covered as platforms, with battlements or plain parapet walls. Some of them indeed, built within that period, we now see finished with pinnacles or spires, which were additions since the modern style of pointed arches prevailed, for, before we meet with none. One of the earliest spires we have any account of, is that of old St. Paul's, finished in the year 1222, it was, I think, of timber covered with lead; but not long after they began to build them of stone, and to finish all their buttresses in the same manner."———Bentham's Ely, p. 40.

circumstances. The idea of the pointed arch seems clearly to have been suggested by the Pyramid, and its origin, must consequently be attributed, not to accident, but to ordination.

But granting for a moment that any of the above mentioned conjectures were true, we should be as far as ever from ascertaining the principles of these edifices. There never was a species of Architecture the properties of which could be determined from the arch alone. Even in the Gothic, where it forms so conspicuous a part, it does not govern in the composition, but follows the general order of things, as it is not a cause, but a concomitant part, and its pointed termination is a consequence arising from a general actuating principle.

Whether the Gothic Architects were the inventors of this Arch, or borrowed the idea of it from others, is not easy to determine; but it is very reasonable to suppose that it originated with themselves, as they were the only scientific builders known to have used the pyramidal figure in the composition of their edifices, except the Egyptians;[*] and it is generally supposed that the latter were ignorant of the art of constructing arches, though, in other respects, an ingenious people. But the Gothic Architects, in using this Arch, did no more, in my opinion, than the Greeks or Romans would have done in similar circumstances. For, if we suppose for a moment, that an Athenian Artist of the age of Pericles, or a Roman Architect of the Augustan age, had been called in to finish a Gothic structure that had neither doors nor windows, he could not, I think, have introduced any other but a Pointed Arch, in an edifice where every part grew up to a point, without being guilty of a direct violation of the laws of art, and of the precepts so strongly inculcated in the Architecture of his own country.

The rule observed by the Gothic Architects, of adapting the form of the arch to the general figure of the edifice, is I believe, consonant with the custom of the ancients, though I cannot find that any writer, ancient or modern, has taken notice of this circumstance. The Grecian temples, which were chiefly of an oblong form, have their doors and windows terminated horizontally, in all the designs that I have seen of the ruins of that country; some exceptions may perhaps be found, but I believe they never occur, except where propriety was made subservient to convenience.

---

[*] The masons of ancient Egypt, though unacquainted with the art of forming an arch, in some will have it, closed both the internal and external apertures of their Pyramids in a manner that resembled, as near as possible, a pointed Arch.— "In the gallery or narrow passages in the great Pyramid of Gize, and in the two rooms of that which is opened at Sacara, the vault over head is formed by the gentle projection of the stones, one above another, till they approach to near a point at the centre."—— Remarks on Prints by Richard Dalton, Esq. p. 54. See the view of the entrance into the great Pyramid, in Sandy's Travels, p. 130.

INTRODUCTION. 5

The Romans, who indulged in a greater variety of forms, furnish us with many examples wherein this principle of uniformity is observed; the doors and windows of their quadrangular edifices being, generally speaking, closed horizontally in the manner of the Grecians, and the apertures of their rotund edifices terminated with semi-circular arches, as we see in the Amphitheatre at Verona, and in the Coliseum, the Theatre of Marcellus, the Temple of Bacchus and Faunus at Rome, &c.* To which uniformity these buildings are indebted for a great part of their beauty. Of this the author of the ingenious Inquiry into the Origin of our Ideas of the Sublime and Beautiful,† appears to have been perfectly sensible. " It is in a kind of artificial infinity;" (he says) " I believe we ought to look for the cause why a rotund has such a noble effect. For in a rotund, whether it be a building or a plantation, you can no where fix a boundary; turn which way you will, the same object still seems to continue, and the imagination has no rest. But the parts must be uniform, as well as circularly disposed, to give this figure its full force; because any difference, whether it be in the disposition, or in the figure, or even in the colour of the parts, is highly prejudicial to the idea of infinity, which every change must check and interrupt, at every alteration commencing a new series."

Indeed, there is no species of Architecture that has the least claim to excellence, wherein this uniform consent of parts, has not been considered as indispensably necessary to the general effect and harmony of the composition. We even find it observed in those simple structures of the Egyptians, that resemble a frustum of a pyramid;‡ which structures probably first suggested the idea of doors with oblique sides, narrower at the top than at the bottom, as described in the sixth chapter of the fourth book of Vitruvius. Doors of this sort, having fitness to recommend them, in a building of that kind, as well as convenience, on account of their shutting themselves, a property they derive from the nature of their figure, were wisely adapted in those artless times, and would be equally proper now, if similar reasons should justify their restoration.§

Upon the whole, if the form of the arches, employed in closing the apertures of a regular edifice, contributes in any degree to the effect and harmony of the composition; it is probable, that the order here assigned to each, appears the most natural, and the most consistent with the rules of fitness and uniformity; that is to say,

A pointed arch, in a pyramidal structure.‖
An horizontal arch, (if the term be allowed) in a square or oblong building;¶
And a semi-circular arch in a rotund.**

---

* See Desgodetz, p. 28, 32, 110, 124, 127, and 139.
The Pantheon, one of the noblest monuments of antiquity, cannot be brought as a fair exception to this rule, on account of the various alterations it has undergone, by rebuilding and repairs, after the greatest part of its original members were destroyed by fire. Vide l'Architettura di Sebastiano Serlio, lib. 3. cap. 4.
† Page 143.
‡ See Fig. 26. plate 1. Introduction. See also Norden's travels, plate 101, vol. 2.
§ The late Mr. Stuart has, not injudiciously, introduced one of these doors at the entrance of the new Chapel of the Royal Hospital at Greenwich, with an obliquity less than taught by Vitruvius.
‖ Like fig. 1, 8, 9, 12, 13, in plate 1. Introduction.
¶ Fig. 15 and 16.   ** Fig. 24.

INTRODUCTION.

Whatever merit thefe different arches may poffefs, abftracted from their properties of ftrength, they fhew it to moft advantage by this arrangement; but, on the contrary, if this arrangement were tranfpofed, the incongruity refulting from fuch a difcordant compofition would be fo apparent, as to deftroy the effect of the whole edifice.

The Gothic Architects appear to have been no ftrangers to the propriety of the preceding ordinance, as is manifeft from the various kinds of arches they employed, to coincide with the contingent forms of their buildings; and whoever fhould undertake to determine their ftyle of Architecture, from the pointed tendency of the arch alone, would not always be correct in his decifion. Their beft Artifts made it a rule, to adopt the arch that was moft congenial in its form to the figure of the edifice. The femi-circle was therefore excluded, becaufe their ftructures were never round: but where the afpect of the edifice was horizontal, the apertures alfo were clofed horizontally.

The Refectory at Batalha,* which is not an inelegant fpecimen of Gothic Art, furnifhes us with a ftriking example of this nature, that enables us to reafon on facts. The general Figure of the plan, and of the elevation, of this Refectory, is nearly like that of a low Grecian Pavilion, and all its apertures, without exception, are clofed in the Grecian manner;† but if each of its buttreffes, inftead of terminating under the cornice, had been carried above the railing and crowned with a pinnacle, and the ends, if finifhed, with fharp pointed gables, its apertures, in that cafe, I conceive, would not have been clofed horizontally, but with a pointed arch, fomewhat fimilar to thofe of King's College Chapel, at Cambridge. Hence it appears, that it was propriety, and not caprice, that influenced the Gothic Architects in clofing the apertures of their edifices; and that a building may be in the true ftyle of the modern Norman Gothic, without poffeffing a fingle pointed arch.‡

Some examples, I am aware, may be pointed out among the Greeks and Romans, of a practice contrary to what is here laid down; but thefe exceptions are merely accidental, or introduced through neceffity, which often excludes all other confiderations.

Among fuch a number of Architects,§ as flourifhed during the continuation of the Greek and Roman empires, there muft, no doubt, have been many, who were ignorant of the true principles of this complicated art;¶ and, from the defigns of fuch perfons, we cannot with propriety draw affirmative conclufions, nor confider their works in any other view, than as monuments of errors; and no errors, however ancient, or however countenanced by long practice, are fit objects of imi-

---

* See the plate of the Weft Elevation of this Refectory.
† See fig. 23, plate 1. Introduction.
‡ Buildings of certain kinds may be in the true M. N. Gothic Style, without pointed Arches, but churches cannot, for reafons which will be fhewn hereafter.
§ Vegetius fays, that in his time 700 Architects were computed at Rome.

¶ Vitruvius, in Proem to B. iii. and B. vi. mentions, that there were many in his time profeffing themfelves Architects, who were ignorant, not only of the found principles of Architecture, but of every thing relating to building.

INTRODUCTION. 7

tation. If rules were laid down for determining, with precifion, what ancient monuments are of the true ftandard principle of correctnefs, they would greatly contribute to accelerate the progrefs of Architecture. But, to afcertain fuch rules, would require the qualifications of the Philofopher, united with thofe of the Artift. He, whofe mind is enlightened by thefe reafoning powers, knows how to ftamp a juft value upon works of real merit, and to reject any excrefcence that " Old Time," as Milton fays, " with his huge drag-net, has conveyed down to us along the ftream of ages."

EXPLANATION OF THE SEVERAL ARCHES
IN PLATE I.—INTRODUCTION.

Fig. 1. THE common pointed arch.*
Fig. 2. An arch of the third point. This arch is ufed in many parts of the church of Batalha. In the fide elevations of churches it has an agreeable appearance, as being an equation between the high pointed and low arch.
Fig. 3. A fection of the common pointed arch. Fig. 1.
Fig. 4. Fig. 5. and Fig. 6. Pointed fcheme arches.
The points from whence the upper parts of Fig. 6. are formed, are found thus: Divide the circumference of one of the circles into fix equal parts; from 1 to 4 of thefe divifions, draw the indefinite right line 1 4 b, and where it interfects the line c b, as at b, drawn perpendicular to the bafe line c 6, will be the point required for defcribing the portion of the arch from 1 to e; the reft is plain by infpection.
Fig. 7. and Fig. 8. Pointed arches, of contrary flexion.
Fig. 9. Segments of an Ellipfis.
Fig. 10. Segments of the Parabola.
Fig. 11. Segments of the Hyperbola.
Fig. 12. Is the Quadratrix of Dinoftratus.
Fig. 13. Segments of the Cycloid.
Fig. 14. Segments of the Catenarian Curve.
Fig. 15. and Fig. 16. Horizontal arches. The former is taken from the Convent of the Church of the bleffed Conception at Beja, where D. Duarte, the fon of John the Firft, the founder of Batalha, is interred. The latter is copied from a Gothic ruin at Evora, both in the Province of Alem-tejo, in Portugal.

* The manner of forming this, and the fubfequent pointed arches, is fhewn by the lines and indexes of the refpective figures.

Fig. 17. This arch, with its lateral and upper refifting arches, may be feen in Weft-minfter Abbey, at the interfection of the nave and tranfept; a fpecimen of Gothic ingenuity, perhaps, not unworthy of notice.*

All the above arches properly belong to that fpecies of Architecture called modern Norman Gothic; but there are fix of them, viz. Figures 9, 10, 11, 12, 13, and 14, the exiftence of which I have not been able to afcertain by antecedent examples. However, I am averfe from fuppofing, that the Gothic Architects were ftrangers to Figures 9, 10, and 11. The arches that fupport the fpire of the Church of St. Nicholas, at Newcaftle, as well as thofe introduced by Sir Chriftopher Wren, in the fpire of St. Dunftan in the Eaft, are, I think, either of the parabolic, or of the hyperbolic fpecies. Thefe arches, whether recommended by ancient practice or not, poffefs many valuable attributes, which, together with their aptitude and fitnefs for Gothic Architecture, invited me to introduce them here.

Fig. 18. The Crefcent, or Moorifque arch.
Fig. 19. A Moorifque pointed arch.
Fig. 20. A Moorifque pointed arch of contrary flexion. Thefe three arches may be feen at the Puerta de los fiete fuelos, and at the Torre de las dos Hermanas.—Alhambra†
Fig. 21. A window in the Arabian ftyle of Architecture, fketched from the Royal Palace at Cintra, near Lifbon.
Fig. 22. Saxon arches.
Fig. 23. The manner in which the Egyptians and Grecians clofed the apertures of their quadrangular edifices.
Fig. 24. The Semi-circular arch often ufed by the Romans, particularly in their rotund buildings.
Fig. 25. The *Dos d'Ane* arch, ufed by the Egyptians, in the vaulting of the galleries, and fepulchral chambers of their pyramids. See Norden's Travels, plate 49 and 50.
Fig. 26. A fragment of an Egyptian building at Carnac; fee Norden's Travels, Vol. I. where many remains of Egyptian edifices are fhewn, with the fides inclining inwards, like the fide of a baftion. Doors narrower at the top than at the bottom, as reprefented in this figure, probably originated with thefe, or fimilar ftructures.
Fig. 30. This is a non-defcript. I fhall call it the ULNAR ARCH. It is generated by the arms of a man extended at full length, with his breaft placed againft a fmooth wall, and marking, with a piece of chalk held in each hand, the revolution of the arms, moving inflexibly upwards, till the hands meet in

INTRODUCTION.

a point vertical with the crown of the head. In the chancel of Elkſtone church, in the county of Gloucester, there is an arch of Saxon workmanſhip, apparently generated after this manner; but ſuch arches ſhould not be employed, I imagine, except in vaults where a number of ribs, ſpringing from the ſides, converge to one centre. The inverted angles at the top ſhould then be concealed by a pendent orb richly ornamented.

How to find the Joints of any Arch formed of Segments of the Ellipſis, Parabola, *or* Hyperbola.

PROBLEM I.   *Of the Ellipſis.*   (Fig. 27. Plate I.)

From the focus point F, of the Ellipſis, draw as many right lines Fc, as there are joints required in the arch, and from G, the other focus point, draw the lines G a, G a, &c. which ſhall cut the former lines at the points c.c. Biſect the ſeveral angles a c e, and you have c b, the joints required.

PROBLEM II.   *Of the Parabola.*   (Fig. 28.)

From the point F, the focus of the Parabola, draw lines Fa, Fa, through as many points (c.c) in the arch as there are joints required, and from the points c.c, where theſe lines interſect the arch, draw cc, cc, &c. parallel to the axis of the ſection GD. The lines bc, bc, which biſect the ſeveral angles e c a, will then be the true joints of the parabolic arch.

PROBLEM III.   *Of the Hyperbola.*   (Fig. 29.)

Let F and G, be the focus points of two oppoſite Hyperbolas given in poſition, from the focus G, draw lines through the ſeveral joints (c.c) of the arch, and through each of the points where they interſect the arch, as at c.c, &c. draw Fa, Fa, proceed, as in the two former problems, to biſect the angles a c G, and you will determine the joints required.

OF DOMES.

THE above problems, eſpecially the two latter, will be found of great uſe, wherever arches of the parabolic or hyperbolic kind are required; their properties are well known to every mathematician, and their utility in conſtruction univerſally admitted, particularly in magnificent domes, and in ſituations where great weight is to be ſuſtained without much lateral reſiſtance: they have alſo the advantage of requiring no centring, or at least not so much as the arches in common use *. The claim of Sir Christopher Wren to the first rank in his profession, depends more, perhaps, upon his knowledge of the properties of these curves, than upon all his other attainments in architecture. Hence he was enabled to design and execute the superb cupola of St. Paul's, one of the noblest specimens of construction existing, and the chef-d'œuvre of this Artist. This edifice, if deprived of that noble feature, would have very little remaining, besides the western portico, that any ingenious Architect of this age would be ambitious to own; and the church of St. Dunstan in the East would have few admirers, were it not for the geometrical ingenuity of its spire.

In that branch of architecture which relates to construction, Sir Christopher Wren may be considered as equal, if not superior, to any Artist that has appeared in Europe these two hundred years; and what contributed not a little to give him this superiority, was his living in an age enlightened by the genius of a Newton, a Leibnitz, a Huyghens, &c. His mind was enlarged by the sublime discoveries of those great philosophers; he was one of the first geometricians of the age †; and if to his knowledge in mathematical science, he had united, in an equal degree, the true principles of the art of design, the world, for the first time, would have seen a complete Architect.

Since the death of that great man, the art of construction has been much neglected in England; and perhaps there are but very few recent specimens of it in Europe, of bold execution, besides the cupola of St. Genevieve at Paris. Yet, it is remarkable, that four of the greatest Architects that have appeared since the restoration of the arts, are chiefly indebted, for their fame in this art, to their knowledge of construction. These were Filippo Brunelleschi‡, Michael Angelo, Jacques Germain Soufflot §, and Sir Christopher Wren. To the first we owe that stupendous machine, the cupola of the church of S. Maria del Fiore, one of the most astonishing and difficult performances in Europe, being in magnitude not inferior to any thing of the kind, perhaps, the ancient world ever saw. The knowledge of construction gradually declined in Italy, by the death of the old Gothic Architects, and finally expired with Arnolfo Lappi, who began this church according to the Gothic rules; this prevented the completion of its cupola for upwards of a century, during which time, there was not to be found in Florence, nor throughout all Italy, any Architect who would undertake to finish it. This arduous task was reserved for the genius of Brunelleschi, who has rendered his name memorable in the history of the arts, by the execution of that cupola; which Lappi, who began the work, would have considered as a simple operation, and would, as M. Felibien * justly observes, have finished it, had he lived, with more ease than Brunelleschi, though he had been possessed of all the rules of the ancient Greek and Roman Architects.

But Filippo Brunelleschi, however great his merit may have been, has no title to the invention of the modern manner of building cupolas; for this we are indebted to Anthemius and Isidorus, the two celebrated Architects who, by order of the Emperor Justinian, built the present Santa Sophia†, at Constantinople; and to prevent its destruction by fire, as had already happened four several times, they employed no combustible materials in its fabrication. From this epoch we may date the origin of cupolas resting upon the four pillars of a square, which square is gradually formed into a circle by pendentives ‡; an idea suggested by the figure of the cross represented in the plans of all Christian churches. The veneration in which this church has long been held, and the advantages which appeared to result from its new mode of construction, are so great, that it has since been imitated by all the nations of Europe §. The Venetians were the first who set the example to the Italians, by erecting the church of Saint Mark, at Venice, upon a similar plan, about the year 973 ‖. Saint Mark's, together with the cathedral of Pisa, built after the same model at the commencement of the eleventh century, probably supplied Brunelleschi with much information in the construction of the cupola of Santa Maria del Fiore, which last Michael Angelo appears to have copied, in the dome of that immense fabric St. Peter's at Rome.

Hence we are enabled to trace through *Gothic vestiges*, the origin of those stately Domes which crown the principal sacred edifices of Europe; a mode of construction, of which there is not a single instance to be found among the remains of the buildings of ancient Rome, or of Greece; nor in those of the Egyptians; nor in the writings of Vitruvius.

The Abbé Winckelmann does not appear to have investigated this subject with his usual care, or he would not have attributed the above invention to the ancient Greeks, without being able to refer to a positive instance to establish the fact; his conclusion to this effect, rests upon the authority of a piece of sculpture which is upon an antique sarcophagus found in the *Villa* Moirani ¶. But granting that this sarcophagus bears the model of a temple, crowned with a sort of cupola, the original may, notwithstanding, have rested upon a circular base; for how is it possible to determine to the contrary, by any external representation in painting or sculpture*?

We may conceive some idea of the difficulty the moderns find in executing any bold design of this nature, by what we collect from the Life of M. Soufflot. This Artist, though one of the best that ever appeared in France, experienced more difficulty in constructing the cupola of the church of St. Genevieve, than in all the edifices that ever rose beneath his direction. Notwithstanding its weight, impulse, and resistance were ascertained, and the whole demonstrated to be perfectly secure by two able mathematicians, (M. Ganthey and the Abbé Bossut) yet this was not sufficient to screen him from the malign criticism of cotemporary Artists, who maintained that the piers, upon which the cupola rested, were inadequate to sustain the incumbent weight. Time and experience, however, proved the reverse; and when he was on the point of completing his project, the detraction of his enemies affected him so sensibly, that it is generally supposed he died of a broken heart.

M. Soufflot appears to have been very intelligent in the Gothic, as well as in the Grecian style of Architecture: to obtain a knowledge of the former, he visited many of the best Gothic buildings in France, of which he made drawings, studied their construction, general proportions, vaults, &c. Hence he acquired much useful information, that materially assisted him in the design and execution of the church of St. Genevieve. Sir William Chambers informs me, that when he was last in France, M. Soufflot shewed him a large collection of drawings he had made from those edifices; at the same time he expressed his admiration at the excellence of the general proportions he discovered in them, and the superior intelligence their Architects possessed in the laws of construction.

In fine, the noblest monuments of Florence, Rome, Paris, and London, bear ample testimony of the great abilities of the above Artists, and in no part of these monuments are their talents more conspicuous than in the construction of the Domes. It is much to be regretted then, that this superior branch of our art should become neglected or unknown, and that more attention is not bestowed on the sound rules, and demonstrative principles, upon which the art of construction is founded †. The study of our Gothic edifices will be found, perhaps, to contribute very much to its restoration; but nothing can compensate for the want of a thorough knowledge of statics, and of conic sections.

---

* The last appears to be, that we are not warranted to conclude, from any existing model, that the ancients used any such termination to their edifices as a cupola resting on a square basement; their round temples, it is true, were often covered with a semicircular vault, to which they gave the name of Tholus, such as that we now see in the Pantheon at Rome; but this form is very different from that which we have adopted from Santa Sophia, as may be seen by comparing the vault of the Pantheon at Rome with the dome of the Augustins' church in the same city.

† " It seems very unaccountable, that the generality of our late Architects " dwell so much on the ornamental, and so slightly pass over the geometrical, " which is the most essential part of Architecture. For instance, can an arch " stand without butment sufficient? If the butment be more than enough, 'tis " an idle expence of materials; if too little, it will fall; and so for any vaulting: and yet, no author hath given a true and universal rule for this; nor " hath considered all the various forms of arches."—Parentalia, 356.

OF SPIRES.

HAVING taken a short review of the origin and progress of Domes, we shall next proceed to give a brief account of the origin of Spires; a subject of all others relating to architecture the least understood. The observations I am about to offer on this subject, are different from any yet presented to the Public; yet I trust to make it appear, that the real intention of Spires, which has never been satisfactorily explained, is such as I am about to submit.

The Spire of Old Saint Paul's, says Mr. Bentham, is one of the earliest we have any account of; it was finished in the year of our Lord 1222, and was in height 520 feet, according to Stowe's account *. The Spire of Salisbury Church is 400 feet high †; and that of Strasburgh, built by the famous Irwin de Steinbach, is 450 feet ‡. No settled proportion seems to have been observed in the dimensions of Spires in general; sometimes the height did not exceed four times the diameter of the base, whilst, at other times, the ratio of the height, to the breadth taken at the base, was as eight to one. We have an example of the last-mentioned proportion, in the Spires built by Hugh Lebergin upon the towers of St. Nicase, the two largest of which were 50 feet high, upon a base of six feet.

Notwithstanding the amazing height to which many Spires were carried, they were constructed so exceedingly slight, that we should be apt to conclude, on reasoning from theory, that they would be inadequate to sustain their own weight. The Spire of Salisbury, for instance, is but seven inches thick; and that of Batalha is about the same thickness, independent of the embossed work, though almost a fourth part of its superficies is perforated. Great care must consequently have been taken, in selecting the materials employed in constructing such light Spires, especially as they are, I believe, in general connected without the aid of iron cramps; for this metal, when exposed to air or moisture, is subject to contract rust, which in time will shiver in pieces as much of the block as it comes in contact with. I am informed, that the stones of the Spire of Batalha are keyed together, by means of dove-tail wedges of pine-wood §: however that may be, it is pretty certain, that the ancients, upon similar occasions, have adopted this expedient. Alberti mentions his having found wedges or cramps of wood in the ancient Roman buildings; and M. le Roy has observed them in the ruins of an ancient temple in the district of Athens. In the ancient Temple of Girgenti in

---

* Wren's Parentalia, p. 271.
† Ibid. p. 305.
‡ Argenville, Vies des Arch. Disc. p. 33. tom. 1.

§ This circumstance I was informed of after leaving Batalha, by an ingenious gentleman in the Portuguese service (Col. Aufrere), who, from his situation, had frequent opportunities of visiting that building.

Sicily, there were found wedges of wood, in good preservation, after a space of more than two thousand years*. Cramps of copper were also used by the ancients in their buildings, which, according to the account of Father Montfaucon, were tempered to an extraordinary hardness †.

With respect to the origin of Spires, it appears very unaccountable, that neither history nor tradition have preserved the least remembrance of it. There must, nevertheless, have been some specious motive for building them; for we can hardly conceive, that appendages so expensive, and difficult of execution, were merely the result of caprice. If we examine the uses to which the sacred edifices wherein they are employed were appropriated in the 12th century, we shall discover a rational cause for crowning them with Spires: namely, the custom of burying in churches, which about this time appears to have become general all over Europe‡. Now, in consequence of this custom, there were united, in the same fabrick, a cemetery and a church §; it was highly proper, therefore, to build every structure intended for this double purpose, in a style of architecture characteristic of its twofold destination. Impressed with these sentiments, the Architects of those times would naturally look back for precedents of a similar nature among the nations of antiquity; the historians of these nations, as well as the remains of their edifices, would have shewn them, that it was invariably the practice of all civilized people, who believed in the immortality of the soul, and did not hold a republican form of government ‖, to raise lofty pyramids over their cemeteries or places of sepulture. The Gothic Architects, in like manner, have adopted that figure to characterise their cemeteries, and, at the same time, preserved the figure of the cross in their ground plan, the better to denote a Christian temple.

Hence the origin of Spires, and the consequent introduction of pinnacles, pyramidal or pointed arches, angular ornaments, &c.; in short, every vertical part of the whole superstructure was henceforth terminated in a point. Indeed it could not be otherwise, consistent with the true principles of design, which invariably

INTRODUCTION.   15

prescribe an harmony between the several parts; and also between these parts and the general configuration; from the whole of which results an unity of appearance, the most certain criterion of its excellence *.

The reason assigned for the origin of Spires will also apply to the pyramids, or round towers, to be found at this day near many of the old churches in Ireland; for it is observable, that, at the time these towers were built, the Architects of that country were unacquainted with the art of raising a Spire over the pillars, at the intersection of the nave and transept. They had recourse, therefore, to an easier but less scientific expedient, by constructing, upon solid bases, those round pyramids which always terminated like the Egyptian obelisks. And notwithstanding all the learned conjectures that have been made respecting the use of these pyramids, we may reasonably conclude, that they were intended to denote cemeteries.

We may conceive how far the Christians of the 13th century were impressed with the propriety of building pyramids over their cemeteries, from the immense elevation they gave to some of them. That of Old St. Paul's, for instance, was loftier than any of the pyramids of Egypt †. And it is worthy of remark, that they were introduced about the time Science began to revive, and recover its long-lost energy; hence they may not be improperly considered, as so many auspicious monuments of the rising greatness, and returning wisdom of Europe.

When we consider the stupendous monuments of Egyptian power which still exist, we cannot avoid reflecting on the vanity of their founders, and pitying the mistaken system of theology that gave rise to them. Yet if we examine our own country, we shall find more pyramids, even in one province, than exist at this day in all Memphis and Sacara. From this circumstance, one might be induced to suppose, that the origin of Spires amongst us, proceeded in some degree from Egyptian ideas grafted upon Christian principles.

Amongst the Egyptians, the pyramid was held to be sacred ‡; by it they expressed the origin of all things. They placed it over their cemeteries, as the Christians do §, to testify the lively and exalted sentiments they entertained of the immortality of the soul. Its form, which is like that of a flame of fire ‖, (whence it is supposed to derive its origin) is typical of the divine spirit of the deceased, ascending, after its separation from the body, to the divine mansions of repose.

Such, we are given to understand, were partly the motives that induced the Egyptians to adopt the pyramid in their sepultures; that figure, however, possesses

---

* Omnia porro pulchritudinis forma unitas est.——St. Augustin.
† The perpendicular height of the great pyramid is 481 feet, which is 39 feet less than the altitude of the Spire of Old St. Paul's, according to Camden's account.——Vide Greaves's Pyramidographia, p. 69, & seq.
‡ The Egyptians called their sepultures eternal mansions; whereas they gave their palaces and houses the title of caravanserais, on account of the short time we sojourn in this life, in comparison to the time our remains repose in the grave.
§ "Le lieu où sont les pyramides, dit le P. Vansleb, qui fit le voyage d' "Egypte en 1672, est un cimetiere, & sans doute un cimetiere de Memphis; "car tous les historiens Arabes nous apprennent que cette ville étoit batie "dans l'endroit où sont les pyramides, & vis-à-vis le vieux Caire."
Encyclopædie.
‖ Sandys's Travels, p. 127. Greaves's Pyramidographia, p. 69.

many interesting properties, which, independent of these motives, might have recommended it to their notice. Experience has evinced, that in point of durability the pyramid is superior to all other figures; it is also a form the most consonant with the principles of opticks; as, on account of some natural imperfections in the visual organ, it is continually obtruding itself on our senses. For instance, a long range of buildings, viewed from either extremity, will appear to incline to a point. A long avenue, a road, or a canal running between two parallel banks, have the same effect. To which we may add, that the pyramid has the property of conveying an idea of elevation beyond the actual limits of the object. Whether these properties were taken into consideration by the Egyptian and Gothic Architects or not, it must be confessed, they have chosen a form that plays very strongly upon the senses, and from it their works derive no inconsiderable part of that effect which fills the mind with ideas of grandeur.

The moderns, as we have already observed, still continue to use pyramids in their churches and sepultures; although no other reason is assigned for this practice, but that it was the custom of our forefathers. We should recollect, however, that Spires were graceful, and well adapted to the general formation of *their* edifices; whereas in *ours* they are quite the reverse. By attempting to imitate the antique style of architecture in our churches, we have fallen into a compound one, which is neither Grecian nor Gothic, but rather a piece of patchwork, made up of the remnants of three different nations. Italy has furnished the ground plan *, Greece the portico, and France the Spire †. The coalition of these heterogeneous parts, cannot with propriety be called Grecian architecture; yet that is the appellation generally given to it. We must allow, however, that there are some churches amongst us, executed in this mixt style, that are not undeserving of praise.

---

* The Latin cross is the form usually given to the plans of our churches.

† The Gothic architecture, with spires and pointed arches, is generally supposed to have originated with the Normans, who, during the eleventh and twelfth centuries, appear to have been great church builders. "The Normans," says the learned Abbé Fleury, "had ruined a great number of churches, and others were suffered to decay, by the false opinions of the end of the world, which was expected exactly in the year of our Lord, One Thousand. When people saw the world still continue after this fatal year, new churches every where began to be built, in the most magnificent style the times would allow; and not only superior to the houses of private persons, but even to those of the greatest lords."——*Mœurs des Chrétiens.*

INTRODUCTION.   17

OF THE GENERAL PROPORTIONS OF

GOTHIC CHURCHES.

FROM the obſervations which at various times I have made on theſe churches, I am led to ſuppoſe that the general configuration, internally, was uſually deſigned agreeable to ſome definite rules, or proportions; notwithſtanding the component quantities were not invariably diſtributed, in every edifice, in the ſame comparative degree of relation, but were modified according to local circumſtances, or the Architect's conception of optical effects. To convey ſome idea of the nature of theſe proportions, I ſhall here ſubmit the reſult of my inquiry, concerning the general interior diſtribution of the Church of Batalha.

The module, or *datum*, taken in this inſtance, is the breadth of the church internally, which we ſhall ſuppoſe equal to AB. (ſee Fig. I. Tranſverſe ſection of the Church of Batalha.) Form a ſquare AEFB, whoſe ſide is equal to AB, and within this ſquare inſcribe a circle; now CD, which is equal to the ſide of a heptagon inſcribed within the given circle, determines the latitudinal diſtance between the axis of the pillars, and conſequently aſcertains the breadth of the nave and ailes. 2dly. From the points C and D, draw CM and DM, parallel to the ſides of the ſquare, and draw the diagonals AF and BE. The magnitude of the pillar being predetermined, according to the laws of ſtaticks, c muſt be its extremity; let o be the axis; form the pillars cdef, and you will aſcertain the diſtance (o, o) between the axis of the ſeveral pillars placed along the nave.

In proportioning the elevated parts, the ſame *datum* is taken as before. The height of the cluſter columns CD, (ſee the Tranſverſe ſection of the Church) is equal to the breadth of the church. The height of the columns of the arcade FE, is two thirds of CD. The radius with which each ſide of the arch over the nave is deſcribed, is two thirds of the ſubtenſe of its baſe; the ſeveral dotted lines ſufficiently explain the reſt.

In order to ſhew that the above proportions coincide with thoſe of the human body, we have given in the ſection the repreſentation of a human figure, whoſe height is ſuppoſed to be equal to the ſuperior part of the cluſter columns CD. According to the ſize of this figure, the hands, if extended, would touch the two ſide walls of the church. The vertex of the arch will be found as far above the head, as the hand can reach when elevated; that is, a cubit, or one fourth of the whole height.

The procedure of the Ancients, in proportioning their edifices, was not diffimilar to the above, if we are to credit what has been afferted by various writers, who have treated on the fymmetry of buildings. The Ionians, as Vitruvius informs us, modelled the columns of the Temple of Apollo Panionios after the fame archetype. From the *Ulnar* Arch (reprefented in Fig. 30, Plate I.) it may be inferred, that the Saxons alfo have had recourfe, in fome degree, to the like expedient. I pretend not to decide on the propriety or impropriety of fuch analogy: however that may be, it is very remarkable, that Architects fo locally remote, and in fuch diftant ages, as thofe of the *Ionians, Saxons,* and *Normans,* fhould have proceeded fo nearly alike, in attempting to affimilate the proportions of their buildings to thofe of the human frame.

Among the various ftructures of the Ancients, which have been tranfmitted to us, there is none that approaches nearer to the form of a Gothic Church, than the Egyptian Hall, defcribed by Vitruvius (B.vi. c.vi.); a fection of which we have given, at Fig. 2, under the Tranfverfe fection of the Church of Batalha. The fimilitude of which we fpeak is obvious at the firft view; but, were the Hall executed entirely in ftone, the refemblance would be ftill more apparent. The arches over the ailes would require lateral refiftance proportionate to their impulfe; buttreffes would neceffarily follow, as at HK; the vault of the nave would alfo require an adequate refiftance; the nature of the defign would immediately fuggeft the idea of reclined fulcra, extending from the walls of the ailes to thofe of the nave, as at LM. Now, fince materials of this fort could not be eafily obtained to form each fulcrum of one entire piece, the conjunction of different ftones, in the form of an arch, is the only fupport that could with propriety be applied; we fhould call it a flying buttrefs. It is true, that a folid wall of a triangular form would anfwer that end, as is feen in the Temple of Peace at Rome; but the fuperiority of the Gothic manner is evidently preferable in every refpect. In fine, were an Egyptian Hall entirely conftructed of the fimple produce of the quarry, the arches and buttreffes confequently introduced, together with the pillars, windows, ailes, and the uncovered paffages over the ailes, would bear a ftriking refemblance to the body of a Gothic Church.

OF DOORS.

THE principal entrances of Gothic Churches are generally ornamented in a very magnificent manner; the Door is ufually placed within a large porch, which porch diminifhes, as it recedes, in a rectilinear direction; the fides of it are often adorned with an affemblage of flender columns, and mouldings of various forts. Here are also seen statues of Kings, Popes, Saints, and Martyrs, with their respective emblems, canopies, pedestals, &c. Frequently we find as many entrances in the west front, as there are porticos in the Church; but the centre Door, which is generally the largest, is seldom opened but on days of procession, or for the reception of some Dignitary of the Church.

How different the practice of the Gothic Architects, in this respect, from that of the Ancients! The latter made the entrances of their Temples large, and approached them by a flight of elevated steps. The former, on the contrary, made the Doors of their Churches comparatively small, and on a level with the surrounding plain. Sometimes, indeed, we even find steps for descending into these Churches, which some imagine to have been appended in consequence of an accumulation of the adjacent earth; but it is evident that in general these steps made a part of the original design. Some instances indeed of elevated entrances to Gothic Churches are to be met with; but they are such, I imagine, as necessarily arise from the obliquity of their situation.

OF WINDOWS.

THERE is no part of Gothic Architecture which admits of more variety, or is susceptible of a greater display of taste and beauty, than the Windows; the manner in which they are usually formed is as follows*. The breadth of the aperture is divided into three, five, or seven equal parts, with a mullion between each. When the Window is of any considerable height, a transom or cross mullion is placed in the middle of it, for the security of the work. The space between the spring of the arch and the summit of the aperture is filled with tracery work, composed chiefly of tre-foils and quatre-foils, and these are sometimes sub-divided into other different figures. In all manner of tracery work, whether simple or complicated, we find that the intersticial vacuities tend to the figure of a plain or curvilinear triangle, and that their circumscribing lines are generated by geometrical rules. The most beautiful sort of tracery, in my opinion, is that in which the several perforations approach to an equal magnitude.

---

* "The Windows of our Gothic buildings in the reign of Henry the Second were long, narrow, sharp-pointed, and usually decorated on the inside and outside with small marble shafts; the order and disposition of the windows varied in some measure according to the stories of which the building consisted; in one of three stories, the uppermost had commonly three windows within the compass of every arch, the centre one being higher than those on each side; the middle tire or story had two within the same space; and the lowest only one window, usually divided by a pillar or mullion, and often ornamented on the top with a trefoil, single rose, or some such simple decoration, which probably gave the hint for branching out the whole head into a variety of tracery and foliage, when the windows came afterwards to be enlarged. The use of painting and stained glass, in our churches, is thought to have begun about this time. This kind of ornament, as it diminished the light, induced the necessity of making an alteration in the windows, either by increasing the number, or enlarging their proportions; though a gloominess, rather than over-much light, seems more proper for such sacred edifices, and "better calculated for recollecting the "thoughts, and fixing pious affections."

Bentham's Ely, p. 39.

The splay of every Window is in proportion to the thickness of the wall: a large splay appears to have been much esteemed; and when the wall is not of dimensions sufficient to admit of it, then a few mouldings are appended to the sides of the archivolt, resting upon a string or belt, or else upon a sort of corbel, formed into a grotesque head. The *apron*, or sill, has nearly the same degree of obliquity as the sides of the window; at the bottom it projects a few inches, where there is a little channel to prevent the rain from recoiling on the wall.

The piers, between the Windows in Churches, are very narrow, in consequence of the breadth of the apertures[*]; and the great splay of their architraves, together with the half pillars of the ailes, occupy nearly the whole of the wall; so that no plain space remains for the reception of pictures. The Ancients, on the contrary, made the piers of their edifices large, and the apertures comparatively small, which is still considered as the *grand style* in Grecian Architecture; as the magnitude of the piers gave them an opportunity of embellishing the inside with pictures and statues. The Gothic Architects, however, have amply compensated for this deficiency, by making the Windows, instead of the walls, the depository of their pictures; and thus, by commuting the canvass for the glass, they obtained one important advantage, that is, a natural back light, a light which has the peculiar property of giving every production of the pencil the greatest possible degree of force and brilliancy. The various colours of these Windows form a happy contrast with the simple white or gray cast of the structure; and, as they obscure the Church in some degree, they diffuse an appearance of solemnity, well adapted to the majesty of the place.

At Batalha, about five o'clock in the evening, when the sun is opposite the great Western Window, the effect of its painted glass is most enchanting. At this hour the Fathers usually assemble in the choir to chant the Evening Service, whilst the myriads of variegated rays, which emanate from this beautiful Window, resemble so many beams of glory playing around them.

It is in vain that we attempt to restore Gothic Architecture, without the admission of stained glass; especially in Churches, where a degree of obscurity is perfectly consonant with the tombs, inscriptions, and other relicks of mortality we behold on every side. If to these we add the solemnity of the Divine Service, the awful silence, and pensive deportment of the congregation, we must admit the propriety of accompanying scenes of this nature with a solemn shade, since it is allowed by all to be more productive of sublime ideas than light. " Our great poet (to speak in the words of a competent judge of these matters) was convinced of this; and indeed so full was he of this idea, so entirely possessed with the power of a well-

---

[*] In the reign of Edward the First, the Windows were greatly enlarged, and divided into several lights by stone mullions, running into various ramifications above, and dividing the head into numerous compartments of different forms, as leaves, open flowers, and other fanciful shapes, and more particularly the great Eastern and Western Windows (which became fashionable about this time,) took up nearly the whole breadth of the nave, and were carried up almost as high as the vaulting; and, being fit off with painted and stained glass of most lively colours, with portraits of kings, saints, martyrs, and confessors, and other historical representations, made a most splendid and glorious appearance. Bentham's Ely, p. 40.

managed darkness, that in describing the appearance of the Deity, amidst that profusion of magnificent images, which the grandeur of his subject provokes him to pour out upon every side, he is far from forgetting the obscurity which surrounds the most incomprehensible of all beings, but *"

" ———— with the majesty of *darkness* round
" Circles his throne."

OF NICHES.

NICHES were used but sparingly by the Gothic Architects. In some of the earliest structures we trace but few instances of them, and these few are chiefly confined to the exterior of the edifice. In the Church of Alcobaça, one of the most ancient in the Gothic style, I do not recollect to have seen a Niche, or a Statue, that was coeval with the original fabrication; and I may add the same of the interior of the Church of Batalha: a strong evidence that statues in those days constituted but a small part of the ornaments of Churches. The plan of these Niches is generally a semi-hexagon; the head terminating with a projecting canopy.

PILLARS.

THE grand effects so universally admired in Gothic Cathedrals, are, I believe, to be attributed to the artful distribution of the Pillars, with their concomitant scenery. Their magnitude, and relative situations, are proportionate to the dimensions of the edifice, and can only be ascertained with precision by the laws of staticks and opticks. The cluster of little shafts, and the intervening mouldings, on the superfice of each Pillar, give them an appearance of instability, and seem superfluous at first sight; but, on tracing their continuity upwards, we find them branching out in various directions, in the ribs of the nave and aile vaults, and converted into archivolts in the sides of the nave. The manner in which these Pillars are placed is rather singular, but well calculated for effect and resistance: the side of each forms an angle of 45 degrees with the collateral wall, by which their repulsion is greater than if placed parallel, in the same ratio the diagonal of a square bears to its side.

* Burke on the Sublime and Beautiful.

If the Architects of thefe Cathedrals were fo illiterate as they are reprefented, it is aftonifhing how they could have afcertained fo accurately the maximum of their pillars, arches, buttreffes, &c. whilft the Moderns, with all their fuppofed improvements in this art, and the affiftance they evidently derive from algebra, have not yet produced any examples in conftruction equal to what the former have left us.

It is not to be fuppofed, indeed, that all the Architects of thofe times were competent to fuch an arduous tafk; therefore the illiterate ones, whofe talents were limited to the practice of their art, availed themfelves of the excellencies of fome approved model, and thus obviated all calculation refpecting the ratio of force and refiftance. This in fome meafure accounts for the fimilarity we find in the breadth of many of thefe edifices, both in this country, and on the continent. We could adduce many inftances in teftimony of this fact, if we admit the relation of their Hiftorians; but, for brevity's fake, I fhall only quote the ftructures meafured by myfelf, which differ but very little in breadth, as appears by the following comparative eftimate:

	Feet	Inch.		Feet	Inch.
Batalha	72	4	Ely Cathedral	72	9
Alcobaça	72	0	Weftminfter Abbey	72	8

In the conftruction of many of our beft Gothic edifices, we find but few large ftones, whence fome writers have imagined that their Architects were unacquainted with mechanifm, but I believe it proceeds partly from the nature of the quarry, and partly from their mode of building. The fplendor of their works confifted in arches; in the fabrication of which, materials of a moderate fize are to be preferred, becaufe large ftones prefs too heavy on the centres, and require much time and labour in forming their intrados to the curvature of the vault. Where thefe objections did not apply, they often ufed blocks which no human force could raife, independent of mechanical aid. Indeed I may venture to affert, that there is not in Europe, a Gothic Church, or Cathedral of note, wherein three or four of the mechanical powers have not been ufed. It is obvious that their arches, particularly thofe of the naves and ailes, are fupported on the principles of the Statiftical Balance; each ftone of which they are compofed is a fruftum of a Wedge. Pullies were ufed in raifing their cornices, pinnacles, &c. and the Lever was neceffarily employed in moving and fixing them. Here we find the application of the Balance, the Wedge, the Pulley, and the Lever, which four powers would have been fufficient to raife either of the feven celebrated ftructures of antiquity.

How far they were converfant with the theory of thefe powers, is a point not eafy to decide. The experience of paft ages has evinced what great things might be effected in mechanicks, independent of fcientific calculations. The obelifks, columns, and enormous granite blocks of the Egyptians, were tranfported,

INTRODUCTION.

as we are aſſured, without any precognition of the complicated movements of Wheels and Pullies. *Joze Zabaglio*, of Rome, never received any inſtructions, he could not even write or read, yet, by the force of his genius alone, he invented machines which, though ſimple in appearance, produced moſt ſurpriſing effects.

In our own days, we have many ſtriking inſtances of ſimilar inventive faculties in men unaſſiſted by erudition. An obſcure workman, of the name of *Caſhman*, deſigned and executed for the Dublin Society (about eight years ago) a flight of winding ſtairs, aſcending in the midſt of a quadrangular hall, without any apparent ſupport whatever, contrary to all preceding examples.

The Equeſtrian Statue of Joſeph I. King of Portugal, one of the moſt magnificent works of the kind in Europe, was entirely caſt by a perſon of the name of *Bartholemeu da Coſta*, who, as I am credibly informed, is not in the leaſt indebted to the theory of any art or ſcience. This ingenious man, after caſting the above Statue, tranſported it from the foundery to the great ſquare of Liſbon, by machines of his own contrivance, and afterwards raiſed it on its lofty pedeſtal, to the admiration of all the Mathematicians of that country.

FRAGMENTS OF GOTHIC ARCHITECTURE.

PLATE II.

Fig.
1. A Wall of irregular ſtones.
2. A Wall of aſhlar work.
3. Manner of ornamenting the ſuperfice of a wall, as ſeen in the interior of Weſtminſter Abbey.
4. Ornaments ſometimes uſed in tombs, and on the ſuperfice of walls.
5. Ornaments copied from an ancient gothic wall at *Evora*.
6. Manner of decorating a gothic Sacrarium, from the original at *Beja*.
7. A plain embattled top.
8. An embattled top with mouldings.
9, 10, 11, and 12. Various ſorts of embattled tops.
13. A Buttreſs.
14. An arched buttreſs.
15. A Niche.
16. Various ſorts of gothic ornaments.
17. A Teſſellated pavement.

Fig.
18. An impoſt corbel.
19. A Window cill.
20. A Fountain, from the original in the royal cloyſter of Batalha. (ſee the letter A in the general plan.)
21. A Patera.
22. A Pinnacle.
23. A Water-ſpout.
24. A Pillaſter.
25. A Reading-deſk, uſed in choirs.
26. Top of a water pipe.
27. A Trefoil.
28. A Quatrefoil.
29. A Capital.
30. A Pulpit, from the original in the tranſept of Batalha.
31. A Mural Pedeſtal.
32. Machicolations.
33. A Loop-hole.

Fig.
34. An Emblem often seen about high altars.
35. An Infulated pedeftal.
36. An Affemblage of mafons tools of the 13th century, taken from ancient fculptures, and records.

Fig.
37. Characters which I found engraved on different parts of the Church of Batalha. I believe thefe characters have been affigned by the Wardens of the fabrick to the workmen, in order to diftinguifh their refpective performances.

FRAGMENTS OF GOTHIC ARCHITECTURE FROM VARIOUS PARTS OF BATALHA.

PLATE III.

Fig.
1. *Cornice* of the maufoleum of the founder—Upper part.
2. ———— in the north front of church.
3. ———— in the weft front of church—Upper part.
4. ———— in royal cloyfter.
5. ———— to the buttreffes of royal cloyfter.
6. ———— in the maufoleum of King Emanuel—Externally.
7. ———— of the buttreffes of centre chapel. (fee the letter G in the general plan.)
8. A *Cornice* with arched modillions—South front of Church.
9 and 10. *Cornices* of the buttreffes of chapter-houfe.
11. A *Cornice* in the maufoleum of King Emanuel.
12 and 13. *Facias* of the circular towers in the maufoleum of King Emanuel.
14 and 15. Single *Belts*; to the buttreffes of weft front of church.
16 and 17. ———— in fouth front of church—Lower part.
18 and 19. ———— to the buttreffes of the chapels.
20. ———— in north front of church—Upper part.
21 and 22. Double *Belts* in the fouth front.
23, 24, and 25. *Belts* to the rere of the chapels.
26. *Copings* to the buttreffes of royal cloyfter.
27. Vertex *Rib* of the vaults—In collateral chapels.
28. A Diagonal *Rib* of the above chapel.
29 and 30. *Ribs* of the vaults in the maufoleum of the founder.

Fig.
31. One of the principal *Ribs* in the vault of chapter-houfe.
32. Panels of the buttreffes of weft front.
33. Plan of the *Mullions* of the large windows over chapels (fee the elevation of the chancel of Batalha). N. B. The tranfoms of the windows are moulded in the fame manner as the mullions.
34. One of the *Mullions* of large window—Weft front of the Church.
35. The principal *Mullion* of the above window.
36. *Mullion* of the upper windows—North front—Upper part.
37. ———— in the lower windows of north front.
38. *Archivolt* of the circular windows of chapter-houfe.
39. Section of a *water channel* in the roof of the maufoleum of the founder.
40. *Bafe* and *Plinth* of the columns at the entrance of chapter-houfe.
41. *Bafe* of the windows of chapter-houfe—near the entrance.
42. *Bafe* to the external pillars of the maufoleum of King Emanuel.
43. *Bafe* and *Plinth* to the internal columns of the above maufoleum.
44. ———— to the pillars of the maufoleum of King John I.
45. *Bafe* to the column of the tranfept chapels.
46. ———— to the columns of the windows—North front.

INTRODUCTION.

RELIGIOUS CUSTOMS OF THE XIIIth CENTURY.

PLATE IV.

Fig.
1. A Crucifix.
2. Emblem of Religion.
3. Chalice and Patine.
4. A Pax.
5. A Mural Crown.
6. A Crown of Thorns.
7. A Chaplet of Roses.
8. A Celestial Crown.
9. Emblem of the Holy Ghost.
10. Palm and Olive Branch.
11. A Cherub.
12. A Star.
13. A Star of another form.
14. A Boate.
15. An Ewer.
16. A Dominican Friar.
17. A Nun.
18. A Censer.
19. A Phial.
20. A Madona.
21. A Labarum.
22. Symbol of Christianity.
23. Symbol of the Trinity.
24. A Latin Cross.
25. A Cross of the Order of Christ.
26. A Wavy Cross.
27. A Greek Cross.
28. A Raguled Cross.
29. A Strepitum—Used in waking the Friars.
30. An Hour Glass.
31. A Memento Mori.
32. A Discipline.
33. A Lamp.
34. A Bell.
35. A Missal, Rosary, and Cushion.
36. A Psalter.
37. Emblem of Saint Peter.
38. The Baptist's Reed.
39. A Candelabrum.
40. A Wax Taper.
41. A Candlestick.
42. A Seat.
43. A Breviary.

Fig.
44. A Chair.
45. An Horizontal Sun-dial.
46. A Lantern.
47. A Pix.
48. A Flagon.
49. A Folding Chair.
50. A Vertical Sun-dial.
51. A Shield.
52. An Altar and Baldachin.
53. A Mitre.
54. A Tiara.
55. A Cardinal's Hat.
56. A Crozier.
57. A Pastoral Staff.
58. A Water-pot.
59. A Tomb with a Figure of a Warrior.
60. A Tumulus.
61. A Cryptical Tomb Stone.
62. A Coffin.
63. A Font and Sprinkler.
64. A Confessional.
65. Holy Lamb.
66. A Coin.
67. Symbol of the four Evangelists.
68. An Oratory.
69. A Mural Tomb.
70. A Canopy.
71. A Flagration.
72. A Sacrarium.
73. A Seal.
74. A Harp.
75. A Shrine.
76. An Organ.
77. A Procession.
78. The five Colours principally made use of in the Church Vestments, viz. white, red, violet, green, and black.
79. An Ancient Gothic Inscription found at Beja *.
80. A Rota or Wheel, used in the Portals of Nunneries.
81. A Friar's Cell.
82. The word Missal illuminated.
83. The Modern Norman Gothic Alphabet.

* This Inscription is read thus: MISERERE MEI DEUS, SECUNDUM MAGNAM MISERICORDIAM TUAM. *Psalms*, Li. v. i.

DESCRIPTION OF THE CHURCHES OF THE PRIMITIVE CHRISTIANS*.

"THE Church was entirely separated from all profane buildings, at a distance from noise, and surrounded on every side with courts, gardens, or buildings dependent on the Church, all shut up within an enclosure of walls. First there was a gate or entry which led into a peristyle, that is to say, a square court, environed with covered galleries, supported by columns like the cloysters of Monasteries. The poor remained under these galleries, and were allowed to beg at the Church-door. In the middle of the court was one or more fountains, to wash the hands and face before prayer: the holy water-vases were now introduced. At the farther end, was a double porch leading into the Saloon or Basilick, which was the body of the Church. I say this porch was double, because part of it was without the Church, and the other within, which the Greeks called *Narthex*. Near the Basilick, on the outside, were at least two buildings, the Baptistery at the entrance, and the Sacristy or Treasury, called *Secretarium*, or *Diaconicum*, at the farther end; the latter was sometimes double. Along the sides of the Church, were often chambers or cells, for the convenience of such as wished to pray or meditate in private; we should call them Chapels."

"The Basilick was divided into three parts, proportionable to its breadth, by two files of pillars, which supported galleries on each side: in the middle was the nave, as we still see in all the old Churches. At the eastern end was the altar, behind which was the Presbytery or Sanctuary, afterwards called the Transept of the Church. Its plan was semicircular, and finished at top like a niche; therefore called in Latin *Concha*, that is, a shell; the recess was called in Greek, the *Absis*. The Christians, perhaps, at first wanted to imitate the sitting of the Sanhedrim of the Jews, where the Judges were seated in a semicircle, and the President in the middle. The Bishop held the same place in the Presbytery, having the Priests on each side of him. His seat was called *Thronos* in Greek, and was more elevated than the others. All the seats together were called in Greek *Synthronos*, in Latin *Consessus*. Sometimes this place was called Tribunal, in Greek *Bema*, because it resembled the tribunals of the Secular Judges in the Basilicks, the Bishop being as it were the magistrate, and the Priests his counsellors. This tribunal was raised, and the Bishop came down from it to approach the altar. The front of the altar was enclosed by an open balustrade, beyond which was another separate place in the nave for the Chanters or Singers, which on this account was called the Choir, in Greek *Chorus*, or Chancel, from the Latin word *Cancelli*. The Chanters were only simple Clerics appointed to that function. At the entrance of the Choir was the *Ambo*, that is, a raised tribune, with steps up to it on both sides, serving to read the public lessons: since it is called the pulpit, desk, or lobby. If there were but one *Ambo*, it was in the middle; but sometimes there were two, that the altar might not be hid. On the right-hand of the Bishop, and consequently on the left-hand of the people, was the pulpit for the Gospel; and on the opposite side that for the Epistle; sometimes there was a third for the Prophecies."

"The altar was a table of marble or porphyry: sometimes it was of massy silver, or even of gold, enriched with precious stones, for nothing was thought too costly to bear the Holy of Holies. The ceremonies still used in the consecration of altars sufficiently express this respect. It was sometimes, however, only of wood, supported on four feet or columns, rich in proportion, and placed over the tomb of some Martyr; for it was customary to assemble, or build Churches by their tombs; or, at least, their bodies were translated to the places where the Churches were erected. Hence, at length it became a rule to consecrate no altar without placing some relics under it. These sepulchres of the Martyrs, were called their *Memories* or *Confessions*: they were under ground, and the way down to them was before the altar. This remained uncovered, except during the time of the sacrifice, only covered with a carpet, and nothing was placed immediately on it. Afterwards it was surrounded with four pillars, supporting a kind of tabernacle which covered the whole altar, and was called *Ciborium*, on account of its shape, which was that of a cup reversed."

* Vide Les Mœurs des Chrétiens, par M. L'Abbé Fleury, § xxxv.

END OF THE INTRODUCTION.

THE

HISTORY AND DESCRIPTION

OF THE

ROYAL MONASTERY OF BATALHA,

WRITTEN ORIGINALLY IN THE PORTUGUESE LANGUAGE

BY

FATHER LUIS DE SOUSA:

AND

TRANSLATED INTO ENGLISH, WITH NOTES,

BY

JAMES MURPHY, *Architect*.

THE

ORIGIN OF THE FOUNDATION

OF THE

ROYAL MONASTERY OF BATALHA.

DON JOHN, the firſt of this name, and the tenth King of Portugal, finding his kingdom invaded, encamped in the plains of Aljubarrota, in the diſtrict of Leiria, accompanied by a few, but faithful and reſolute ſubjects. His adverſary, another King named John, and alſo the firſt of that name in the regal line of Caſtile, was drawn up in his front, with all the forces of his kingdom, among whom were a great number of Portugueſe, who followed him either through motives of intereſt, or from a miſtaken idea of the juſtice of his cauſe: Matters having arrived to this criſis, a battle became inevitable.

Notwithſtanding the uncertainty of ſucceſs in war, even when both ſides are equal, and the great danger which threatened the Portugueſe, on account of the inferiority of their number compared to that of the enemy, whoſe multitudes covered mountains and valleys; yet our King, finding that he was purſued within his own dominions, could not avoid meeting his antagoniſt, without great diſcredit, if not total loſs of reputation. At the time he reſolved to give battle, he implored the victory of *Him*, who alone has the diſpoſal of it, whence he is called *the Lord of Hoſts*. He alſo invoked the mediation of the Virgin Mary, becauſe the battle was on the eve of her glorious aſſumption; and made a vow, if he came off victorious, to build a magnificent Monaſtery in honour of her. The Lord was pleaſed to crown his

A

arms with fuccefs, notwithftanding the confidence the enemy placed in the fuperiority of their numbers *.

In confequence of his victory, the whole kingdom was fhortly reduced to obedience; but the time which was occupied in different arrangements did not prevent the King from difcharging the obligations of his vow. Though employed in arms, he examined defigns, confulted architects, and fought for artificers. On the one hand, he reduced fome places that held out againft him; on the other, he proceeded to raife this facred Edifice; and thus the work of the Monaftery went on for the fpace of three years. When engaged in the fiege of the *Caftle of Melgaço*, he promifed to give the Monaftery to the order of *S. Dominick*, as expreffed in his will, made many years after in the following words.

"Whereas we promifed on the day we had the battle with the King of Caftile, if the Lord would render our arms victorious, that we fhould order a Monaftery to be built in honour of our bleffed lady S. Mary, on the eve of whofe affumption the battle was fought. After the commencement of the faid Monaftery, *Doctor John das Regas*, of our council, and *F. Laurenço Lamprea*, our confeffor, being with us at the fiege of *Melgaço*, requefted, that we fhould command it to be of the order of S. Dominick: but having fome doubts on that head, becaufe our promife was to build it in honour of our lady, the bleffed Virgin Mary; they anfwered, that the faid lady was much attached to this order, and declared to us for what reafon. Having duly confidered the fame, we confented, and caufed to be ordained, that the faid Monaftery be of the Dominican order."

As foon as the King made himfelf mafter of *Melgaço*, and was returning home, he ftopped at the city of Oporto, and thence iffued his letter of donation to this order, in the beginning of the year 1388.

Father Raymunda de Capua, confeffor to *Santa Caterina de Sena*, was at this time mafter general of the order: In the fchifm which happened in the church, on the death of *Pope Gregory* XI, he took part with the true fucceffor of Saint Peter, and vicar of Chrift, *Urban* VI. This General, finding the province had accepted the convent, confirmed the fame, and the order took poffeffion of it im-

---

* The following account is given of this battle, by *Emanuel de Faria*, in his Hiftory of Portugal:

"The king of Portugal, underftanding the approach of the Caftilians, drew together his forces from *Coimbra*, *Oporto*, and other places, and marched out of *Guimaraens* to give them battle. On the morning of the 14th of Auguft, 1385, he entered the plains of *Aljubarrota*, where he knighted feveral gentlemen. The Caftilians at firft intended to march directly to Lifbon, yet, after fome confultation, they refolved to engage. The forces on both fides were very unequal: the Caftilians are reported to have been 33,000 ftrong, and the Portuguefe but 6,500, befides having fome local difadvantages. The fun was fetting when thefe two unequal armies engaged. The Caftilians, at the firft charge, broke the Portuguefe vanguard; but the king coming up, his voice and example fo animated his men, that in lefs than an hour the multitudinous enemy were put to the rout. The king of Caftile, who leaded his troops, being troubled with an ague, was forced to take horfe to fave himfelf; moft of the Portuguefe, who fided with Caftile, and were in the front of the army, were put to the fword, for no quarter was given them. The Royal Standard of Caftile was taken; but, many pretending to the honour, it could not be decided by whom. The number of the flain is not exactly known, though very great on the part of the Caftilians: of their cavalry 3000 are fuppofed to have perifhed, and many perfons of diftinction. This is the famous battle of *Aljubarrota*, fo called, becaufe it was fought near a village of that name.

mediately. *Father John Martins*, profeſſor of theology, and a perſon of great repute throughout the whole kingdom for virtue and learning, was ordered to preſide over it. The work being conducted under royal agents, he, with his companions, did nothing more than ſay maſs, preach on feſtival days, and reſide as in their own houſe. But they had neither the directing, nor planning of any part, becauſe the whole ſtructure was on the King's account, and carried on by thoſe, who in his name preſided there.

Of the ſituation the King choſe for the Monaſtery, and the reaſons which induced him to build it there.

THE King, deſirous of building a Temple and Monaſtery, which ſhould ſurpaſs the moſt ſtupendous, not only in Spain, but throughout all Chriſtendom, ſucceeded in realizing what he conceived in imagination; for neither *his* age, nor many ſucceeding years, witneſſed ſo grand and magnificent, ſo perfect and elegant an edifice. He invited from diſtant countries, the moſt celebrated architects that could be found, and collected from all parts, the moſt dexterous and ſkilful ſtone-cutters: to ſome he held out honours, to ſome great wages, and to others both. The ſame of the greatneſs of the ſtructure, drew from all parts of the kingdom multitudes of workmen, for works of the kind are attended with this good, they maintain numbers of poor people. The King being poſſeſſed of enormous wealth and having faithful overſeers, the work ſeemed to aſcend like an exhalation.

But before we enter upon the particulars of the fabrick, it may not be amiſs to ſay a few words, concerning the motives which induced this prudent King, to conceive in his exalted mind, the idea of raiſing a pile, the admiration of the world, in a depopulated deſert; deſtitute of ſhady woods and cooling ſprings, and in a low, humid ſituation. In great cities and towns, or near them, are to be found many perſons of diſcernment, to praiſe and eſtimate every thing that is meritorious. Wood and water are uſeful and ornamental accompaniments to a great convent: an elevated ſituation adds to its dignity, extends its proſpect, and contributes to the health of its inhabitants. On the contrary, a low ſituation conceals its grandeur, diminiſhes its appearance, and, in conſequence of its generating infirmities, renders it a living ſepulture.

Notwithſtanding theſe weighty objections, the King, agreeably to his previous reſolution, would not change the ſituation in which he received the Divine favour, as declared in the words of his teſtament. Now, ſince it was to be built where the battle commenced (in which ſituation he immediately commanded an oratory to be erected to *St. George*), or in its neighbourhood, there was no place in the diſtrict better adapted for the purpoſe than that which is called *Canoeira*; the ſoil about it being dry, and a fine river flowing throughout the year, very ſerviceable to the Monaſtery. And a little lower down, the eye is ſtruck with extenſive and fertile plains, watered by its courſe, and by that of another great river. Theſe circumſtances are of great conſideration; for where there is water and ingenious people, there can be no want of refrigerating ſcenes. With reſpect to the lowneſs of the ſituation, it is compenſated by the loftineſs and magnitude of the work, which, the King imagined, would draw thither ſufficient inhabitants to form a reſpectable neighbourhood. As to the humidity of the ſite, the Architects aſſured him, it would dry with the edifice, or at leaſt ſuch parts of it, as were injurious to health; and they were not miſtaken in this reſpect, for, although the Monaſtery experiences moiſture at times, it is not unhealthy. Moreover, the ſituation is not more than half a league from St. George, where the engagement began. All theſe conſiderations outweighed the other objections.

To convey an appropriate idea of this ſtructure, is, more properly, the work of the pencil, than the pen; better adapted for the canvaſs, than the deſcriptive page; becauſe all deſcription muſt fall ſhort of its excellence: it being impoſſible to particularize in writing every minutiæ, which is very eaſy in the language of light and ſhade. The hiſtorian can give but a general idea of things, whereas the painter, by the magick of his art, repreſents every part however ſmall.

As a proof of this, it happened that ſome ſtrangers of good taſte, who had intimation of this ſtructure in their own country, from the copious narrations of our fathers, were greatly ſurpriſed upon ſeeing it, to find how far it exceeded what fame reported; yet, theſe were men who had ſeen and compared the firſt edifices in Europe. With this apology, we ſhall proceed to give as clear and accurate a deſcription of it as we are able.

Deſcription of the Church internally, ---- Dimenſions and Properties of the Edifice.

THE firſt appellation the King gave the Convent, was *Canoeira*, on account of its being within half a league of a ſmall village of that name. The title by which it is known at preſent, derived from the cauſe of its eſtabliſhment, is *Batalha*. Our ancient fathers, more religious than claſſical, call it improperly *De Bello:* a name which would be proper and applicable, were we to take its ſignification from the Latin adjective *bellum* fine, or beautiful, inſtead of the ſubſtantive *bellum*, which imports war.

The Church commenced with amazing grandeur: the workmen, conceiving an idea of its elevation, from the vaſt extent of the foundation, thought it impoſſible to find men and treaſure ſufficient to complete it. The body of the Church alone, from the principal entrance (which is at the Weſt, and runs towards the Eaſt, agreeably to the manner of the Ancient Churches), is 300 palms * long, to the firſt

* A Palmo Craveiro of Portugal is 8 ⁴⁄₅, inches Engliſh meaſure. Or, as 43 is to 40, ſo is a Portugueſe palm to an Engliſh foot nearly.

OF THE ROYAL MONASTERY OF BATALHA.

step of the great chapel; and thence to the wall at the back of this chapel 60 palms, making the whole 360. The breadth is 100 palms, equal to one third of the length taken to the first step of the great chapel.

The above dimensions, correspond with the elevation, agreeably to the just proportions of Architecture. The height is such, that an athletick slinger can scarce cast a stone to the vault of the nave; for it is 146 palms from the pavement of the Church*. Of the two ailes and the nave, the latter is 33 palms wide, and the former 21½ each. These, added together, fall short of 100 palms, the breadth of the Church, as already mentioned; but the deficiency is made up by the addition of the pillars, of which there are eight at each side, of twelve palms diameter at the base.

The nave and ailes are covered with vaults, which, together with the pillars and walls, are all of hewn stone, united with such accuracy, that the joints are scarce perceptible. The thickness of the walls is equal to that of the base of the pillars, that is, 12 palms. The workmanship wants only to be polished to exhibit all the elegance that art could devise. The quality of the stone is the same throughout the building, nor could all Spain produce better for similar purposes: for, notwithstanding the colour is very white, yet it is solid and durable, and at the same time yielding to the chissel. This is sufficiently evident from its having already stood 200 years, † without any traces of decay, with the exception of the loss of its original whiteness: it is not so tarnished, however, as to lose every trace of its primitive gracefulness. In this respect, it may be compared to a beautiful face, exposed to the influence of the sun and air, yet is scarce ever so injured as to lose every trace of its former charms. Thus it is with this stone, which derives from antiquity a tint, neither brown, black, nor disagreeable ‡.

The Transept is 30 palms wide, just the fifth part of its length, (that is, 150 palms). The walls of the body of the Church are all plain, not interrupted, nor excavated (as is often seen in others) by a number of chapels, except, that at the right hand of the principal entrance there is a large arch, leading to a beautiful square room, of which we shall speak hereafter.

The front of the Transept, at each side of the high altar, is subdivided into four chapels, that is, two at each side §. The first, next the Sacristy, is dedicated to *S. Barbara*: this chapel contains a low Sepulture of a Cardinal, whose name and family are no longer known; it is pretty certain, however, that he was of the blood royal. The second is dedicated to our *Lady of the Rosary*;

---

* See the Transverse Section of the Church.

† The Historian appears to have written his account of this structure near 200 years ago, for the building is now about 400 years old.

‡ Internally it retains its natural colour, but the outside of the Edifice has contracted a yellow lamina, which gives it a very picturesque appearance.

§ See the letters H. L. J. K. in the General Plan. See also the *Elevation of the Church.*

B

here we see a high monument of good workmanship, to which *King Afonso V.* ordered his *Queen Dona Isabel* to be translated, who died at *Evora*, in the year 1455. The third, which is at the right hand of the high altar, is called the chapel of our Lady *da Piedada* i. e. of Mercy; here the remains of King John the Second are deposited*. The fourth chapel, the founder of this structure, appointed for the remains of the Grand Master of the Order of Christ, *Don Lopo Diaz de Sousa;* a situation his valour and great services well merited. *Count de Miranda Aurique de Sousa,* now the successor and heir of this Master, deposited here, in our days, the remains of his consort *Dona Mecia.*

In the middle of the great chapel, below the steps of the altar, King *Don Duarte* and his Queen *Dona Lianor* are inhumed, in a large tomb of the same kind of marble as that of the edifice. This tomb is without any inscription, and distinguished only by the cumbent effigies of both, with their right hands joined: his left rests upon an escutcheon of arms, that of the Queen holds a book. The effigies are of excellent workmanship, and said to be executed after nature. Opposite the Transept entrance, at the end of the cross, is the chapel of our Saviour, with a large and beautiful stone altar-piece of modern workmanship †. The other five chapels, that is to say, the great and the four collateral ones, may be said to have no altar-pieces: for, allowing that the great chapel and that of the *Rosary* have altar-pieces, they are too small, and of very indifferent workmanship; which clearly evince, that they are not of a piece with the rest of the fabrick, nor correspondent to the intention of the founder. As to the other three, they are entirely destitute of altar-pieces ‡. There are windows in each chapel, in the situation wherein the altar-pieces should properly have been placed; whence we may infer, that, were it the intention of the founder to make altar-pieces of wood, or of stone, he would have made them at the beginning, or have left a vacancy for inserting them afterwards. In my opinion, he intended, according to his wonted magnificence, to make them of silver, formed like the effigies of Saints, not fixed, but moveable, so that upon festival days, they might be brought to cover the altars; and, I am the more convinced of this, by his having bestowed to the treasury fifteen busts, as we shall see hereafter.

In each of the five chapels, the windows are richly illuminated with emblematic paintings on various devout subjects. And though the situation is bleak in consequence of the great height of the walls, yet, the greater part of the glass is still entire, and the original casements are not injured in the least. There are persons residing in the house, who have fixed salaries for keeping them in repair.

---

* The body of this King, who is said to have died in consequence of poison, was conducted from the Cathedral of Silves, to this Monastery, where it remains to this day incorrupted.

† See the letter F in the General Plan.—The architecture of this altar is in the style of Borumini, a style well known for its want of taste; this, together with its being a modern appendage, induced me to think it not worth representing in the Longitudinal Section of the Church.

‡ These chapels are at present supplied with decent altar-pieces.

The great chapel is illumined with fourteen windows, of which ten are adjoining the altar-piece, that is, five below, and five above. The height of each is 42 palms, by 3½ in breadth. Consequently the aperture of every window is 147 square palms, which aperture is filled with stained glass, without either transom or mullion.

The other four are placed two on each side, and so high, that they receive their light over the collateral chapels; these are 20 palms high, and 12 broad. To secure the glass, each has two mullions of one palm in thickness, so that we may compute each of these windows at 200 palms of glass.

The four collateral chapels have each three windows, which differ but very little in size; upon an average, we may compute the height of each at 40 palms, by 3 in breadth, with the same quantity of glass.

Description of the Mausoleum of the Founder.

WE mentioned before, that on entering at the principal door of the Church, there is an arch at the right-hand, inside of which is seen a quadrangular room, that measures 90 palms on every side. The stones of which it is composed, are of the same quality with those of the Church; it is vaulted, and the centre crowned with an octangular lanthorn, supported by eight pillars; a contrivance which gives light to the inside, and support to the arches of the ailes, whilst it adds to the magnificence of the chapel. The distance from one column to the other, measured on the diameter of the octagon, is 38 palms. The whole is decorated with beautiful windows, which, like those of the Church, are ornamented with stained glass, bearing the arms of the kingdom, and the emblems of the King as he commanded them to be made. As the lanthorn rises considerably above the first tier of windows, there is a fascia of stone at a certain height to bind the work, and upon this fascia are placed other windows, directly in a line with those below and similar in workmanship. The whole height, from the key-stone of the vault which covers the lanthorn, to the pavement of the chapel, is 92 palms.

The above lanthorn, forms a kind of pavilion to a sepulture and an altar, which are placed between the pillars; the sepulture the King had made for himself, and for Queen Philippa * his consort; though the Church was built

---

* Queen Philippa, was daughter to the Duke of *Lancaster*, who, at the instigation of King John the First of Portugal, asserted his right to the Crown of Castile, to which he appeared to have had a legal title by his wife *Constance*. Upon this presumption the Duke set sail from Plymouth, and arrived at *Corunna* in *Gallicia* the 25th July 1386, and landed with 2,000 horse, and 3,000 archers, besides other forces; several persons of distinction accompanying him.

The Duke, at this time, was sixty years of age, without any grey hairs; his person was tall and well shaped, he was affable and modest in conversation, and in all respects answered his royal descent. With him came his wife Constance, and his two daughters, *Philippa* by his first wife, and *Catherine* by the second. Scarce was he landed at *Corunna*, when that place acknowledged him as lawful Sovereign, as did also the city of *Santiago*, and the greatest part of the kingdom of *Gallicia*.

When the Duke landed in Spain, King John was at *Leyria*, and having at his sole expence, yet, he refused with his usual greatness of mind, to occupy the best place in it. The monument is of very white marble, and ornamented, on every side, with foliage of briars in demi-relief, bearing thorns and berries: at intervals are these French words, *Il me plait pour bien**.

The meaning of this emblem is so exalted, that we must allow this Prince to have great understanding; because, if we take the true signification of the Latin name of a briar, which is *Rubus*, it means a bush, or bramble. This is an allusion to Moses, who was called by the Lord, amidst his people to lead them forth from bondage. The King being called in like manner, to a similar enterprise, which, like Moses, he obeyed without hesitation, with the words, *Il me plait*; as much as to say, that in compliance with the divine call, he would most cheerfully undergo every difficulty and labour for the preservation of his people.

If we consider it an allusion to the mysterious branch of the sacred text †, which is likewise a kind of bush, or bramble, the device will be very applicable; for, our Prince was confessedly another Abimelech, as to what regards his birth and beginning, but by means of works of valour and virtue, contrary to him, he finished his days full of prosperity. Abimelech, in order to reign alone, treacherously put to death seventy brothers, lawful children of his father, being himself illegitimate. Our Prince, on the contrary, was so void of ambition, that he acknowledged his two brothers who absented themselves, to be nearer and more worthy heirs to the crown than himself; therefore, he intended nothing more, than to preserve it for them under the title of Defender. That of King he did not assume, till the united voice of the people, and the absence of his brothers compelled him to it. If then Abimelech was a fire that came out of the branch, which burnt his city, his people, and himself; our King was a fire, or luminary of honours, a victorious hero, and an ornament to the illustrious throne of Portugal. After reigning many years, full of prosperity, rich and contented, he departed in peace, surrounded by children and grand children, and was beloved by his people, as much as Abimelech was detested. So that the device, taken altogether, is sententious, and highly applicable to the Royal founder.

Upon this tomb are two cumbent figures of marble, one of the King in

---

agreed to meet the Duke at *Ponte-Mouro*, set out with a numerous retinue. They met, upon the 1st of November, in a plain near *Melgaço*: here it was agreed, that, if the Duke succeeded in his enterprise, he should give twelve towns with their territories to the King of Portugal, as a dowery with his daughter *Philippa*. The Princess, accordingly, was conducted to him, and they were solemnly married upon Candlemas-day, in the year 1387. Immediately the Queen's household was established, and a splendid revenue assigned her.

The King having spent two months with his Queen at *Oporto*, marched at the head of 3000 Spearmen, 2000 Cross-bow men, and 5000 Infantry, to meet his father in law at *Bragança*. They entered the dominions of Castile, and took *Cosira*, *Carbe*, *Mantilla*, *Rofales*, *Valderas*, and *Villalobos*. Though *Gallicia* had received the Duke as lawful King, yet, no place in Castile admitted him but by constraint. Hereupon the King told him, that, to make an absolute conquest, he had better return to England for more forces. The Duke approved his advice, and they returned to Portugal, where Ambassadors came from the King of Castile, offering, that Prince Henry, heir to the crown, should marry Catherine the Duke's daughter. The Duke assented, and the war betwixt him and Castile ended. Soon after he returned to England.

Du Farid's History of Portugal.

* See No. 28 in the Plate of *Ornaments, Mottos, &c.*

† Judges, c. 12.

complete armour, the other of the Queen on his right, with their right hands locked in each other *; their heads are turned towards the weſt, and each has a particular inſcription carved on the ſide of the tomb, which, being large, ſhall have, a ſeparate chapter. The altar above mentioned is at the foot of the ſepulchre, adjoining the pillars which ſupport the lanthorn, in ſuch a manner that the altar and ſepulchre together make but one chapel.

In the wall oppoſite to the entrance, at the King's right-hand, are four mural ſepulchres under as many arches of exquiſite workmanſhip; upon the tablet of which ſepulchres are repreſented eſcutcheons of arms, mottos and devices in mezzo relievo, relating to the King's four ſons who are interred therein †.

Don Pedro.

IN the firſt ſepulchre is interred the Infante *Don Pedro*, the eldeſt of the four; he was Duke of *Coimbra* and *Monte mor*, and governor of the kingdom for eleven years during the minority of King *Afonſo* V. his nephew and ſon in law. We might truly affirm that he was the moſt perfect and juſt governor the country experienced for many years. It is related that this Infante viſited ſeven parts of the world ‡: however that was, there is no doubt but he travelled a great deal; in Germany he was preſent with the Emperor Sigiſmund upon ſome memorable occurrences. His death was unworthy of the great virtues with which he was endued: he was unfortunately killed in a battle, called in ancient records *Alferroubeira*, in which he was the ſole object of the enemy's purſuit, and almoſt the only one that fell a victim to his rage, though of all others the moſt deſerving of life.

His ſepulchre exhibits the device of the Order of the Garter, with its motto, of which he was a Knight. This order belongs to the Kings of England, and they confer it as the higheſt mark of diſtinction on their principal friends, and other meritorious perſons. In another part of his ſepulchre are to be ſeen balances, mixed with branches, from which acorns are ſuſpended like thoſe of the ilex, with this motto, *Deſir* §. It is ſaid that the balances are a particular device the Infante had of the great Archangel S. Michael, in a certain miracle attributed to his birth; a device, which coincides with the actions of one who held the adminiſtration of the republic, and with the ſolemn

---

* See the Plate of the Effigies of King John and Queen Philippa.

† Theſe are the King's four ſons after the Hereditary Prince *Don Duarte*, who ſucceeded him to the throne, and for whom he left the great chapel. We do not reckon the Infante *Don Afonſo*, the firſt child, who died in his infancy, and is interred in the Cathedral of *Braga*.

‡ Don Pedro ſet out from Portugal in 1424 with a train ſuitable to his quality, and travelled over a great part of Europe, Aſia, and Africa, in which employment he ſpent four years. As travels at that time were very rare, eſpecially among perſons of his rank, his gave riſe to many fabulous reports, which have ſince rendered the truth itſelf ſuſpected.

§ No. 29 in Plate of Ornaments, Mottos, &c.

promise he made to preserve justice; but as serious persons seldom make great promises, he assures us of his good intentions only, by the word *Desir*, which means, *I wish*.

As concisenefs is commendable in all mottos, his must be admired; for, though it confifts but of two fyllables, it fufficiently fatisfies us with the whole of the device, which is fuch as might be expected from an excellent governor. His conduct in this refpect appears to be a true imitation of the great Sages of Greece, who exchanged the arrogant name of Wife-men, (for in reality none but God is wife), for that of Philofophers, or Lovers of Wifdom. The Infante in like manner modeftly promifed, not actions, but fincere intentions, in the public government; and conformably to this promife he performed excellent things indeed; like another Solomon, whofe ardent love of knowledge obtained of Heaven the greateft earthly wifdom *.

As it is natural for people to wifh for the ancient times and golden ages, when they had abundance of every kind of fruit, and lived in perpetual peace without oppreffion; it appears, the branch with the acorns indicates, that he would endeavour to introduce by his good government a fimilar age: becaufe, this fruit is fuch, that, without any care, or labour, the earth produces it, and people had only to gather it, and thus they maintained themfelves in thefe primitive ages, as Hiftorians inform us.

We may alfo add, that it was a cuftom among the ancient Romans, to honour, with a crown of oak, the citizen who faved the life of another in battle †. This they called the *Corona Civica*; which was fo much efteemed, that, after the commencement of the Emperors, they accepted it as the infignia of clemency, in preference to a crown of gold or of any other kind ‡. It is faid, Auguftus conferred fuch with his own hands, and permitted himfelf to be crowned with one, confidering it a gift conferred upon him by the human race, for the general peace he gave to all the world. Now as the Infante was in the kingdom fuch a great perfonage, he always wifhed to make it appear that he governed with juftice and equity between citizen and citizen, by which means, he promoted the good of all: hence we may conclude, confidering every thing, that the device was judicioufly adopted.

Don Anrique.

THE fecond fepulchre is occupied by the next in age, the Infante *Don Anrique*, Duke of *Vizeu*, Lord of *Covilham*, and Mafter of the Order of

---

* 1 Kings c. iii. † Pliny, lib. xvi. c. 4. ‡ Ibid. lib. xvi. c. 4.

Chrift. It is faid, that he was elected King of Cyprus; indeed it appears by the effigy over his fepulchre, that he was dignified with a royal crown. However that was, we know for certain, that his foul was crowned with many and exalted virtues. He led in perpetual continence, a retired and philofophical life, cultivating all the ufeful fciences, and, in particular, thofe of Cofmography and Geography, by which he opened the way, to the firft difcoveries, of the feas and unknown lands of the coaft of Africa. For this purpofe he lived at Sagres, in the kingdom of Algarve, in a fmall village, which, at prefent, is called after him *Villa da Infante*. In thefe fublime purfuits, he was bleffed with a long and tranquil life, and a peaceful death.

His efcutcheon bears the device of the Garter, with which he was invefted in his early days, becaufe he was related to the King of England; and another efcutcheon bears the Crofs of the Order of Chrift. Among the fculptures are to be feen little ftalks, with tender branches fhooting out of them; the form and fruit of which refemble the evergreen oak, with its acorns; the branches are fhort and twifted, and the leaves fet with fharp points. He, who purpofed to cultivate the barren deferts of Lybia, with infinite danger by fea and land, which at firft was the object of his difcoveries, and the beginning towards humanizing that barbarous country, and giving it the knowledge of the true God, might well exprefs his good intention, and the difficulty attending the enterprize, in the hardinefs of an oak, and by the dry fruit which it yields. His motto is alfo expreffive of this undertaking, *Talant de bien faire* *, that is to fay, *A difpofition to do good*.

Though this expedition put him to great expence and trouble, he never entertained a thought, that it would be attended with more advantage, than the oak and its fruit are to the mountains; as may be inferred from a book which he caufed to be written on the progrefs of thefe difcoveries, wherein he conveys the fame thought under other emblems, very ancient and pointed in their fignification; namely, the Pyramids, the works of the ancient Kings of Egypt, which, though raifed with immenfe expence and labour, and reckoned among the wonders of antiquity, yet ferve for no human purpofe. However, they clearly exprefs the meaning of the Infante in his writings. The book we allude to was fent by the Infante to one of the Kings of Naples, and it was feen in the city of Valencia, in Arragon, in the Cabinet of the Duke de *Calabria*, (the laft defcendent in the male line of that King), who went to Naples, and ended his days with the title and command of Viceroy.

* See No. 30.—Plate of Ornaments, Mottos, &c.

Don John.

NEXT follows the Infante *Don John*, Master of the Order of *Santiago*, and Lord High Constable of *Portugal*. This Prince married the daughter of the Duke de *Bragança*, grand-daughter of the Lord High Constable *Don Nuno Alvres Pereira*. *Don Afonso*, the brother of the latter, had two daughters, from whom most of the Kings and Princes of Europe participate of the blood of this illustrious family.

His device consists of extended branches, with wild strawberries, amongst which are suspended square pouches, like those formerly used, with three shells over each. The motto is in French, like those of his father and brothers, because, in his time, the French language was much esteemed and current among Princes for its courtesy and politeness: it is this, *Je ai bien reson**, *I have good reason*.

As we are unacquainted with any particular actions of this Prince, we are at a loss to know, what his motive is grounded on, for adopting the emblem of the wild strawberries, which no doubt was very proper. The only reason we can assign for it, is his devotion to the Glorious Baptist, as may be collected from his altar, which, together with his name being John, induced him, in all probability, to adopt the emblem of a tree bearing wild fruit. We know the great Saint never looked for any better food; and it is not inconsistent to assimilate two Saints, the one by devotion, the other by obligation. Now, if the fruit of the forest denotes the Baptist, the pouches and shells are devices of St. James, of whose Order the Infante was Master.

Don Fernando.

THE last sepulchre † is the place of the last and fourth brother, the Holy Infante *Don Fernando*, Master of the Order of *Avis*, and the sixth son of King John. His escutcheon bears the Royal Cinques upon a Cross, adorned with the flowers of his Order. The emblems on its field, consist of branches like those on the sepulchre of the Infante Don John, but with this difference, the latter are expanded, whereas the former are wreathed, and bear a different fruit. Some have supposed these circular branches to be of the thorn tree, in consequence of the resemblance they bear to a crown;

---

* No. 31.—Plate of Ornaments, Mottos, &c.
† Sepulchre. So called by our Historian, though I conceive it might more ‖ properly be named *Cenotaph*, as this Prince was taken prisoner by the Moors in Africa, where he ended his days in slavery, and is interred.

but as they have no thorns, we cannot readily admit the suppofition. The emblem, however, in this fenfe, would be excellent, and altogether prophetic of the thorns and troubles which he afterwards experienced among the Moors.

As he was a Saint, and loved the thorny crown of Chrift, perhaps he was too modeft to declare to the world what he forefaw, that it might not appear he boafted of virtues before hand, which his actions and patient fufferings afterwards evinced he poffeffed; the thorns therefore are wanting to complete the emblem. Although we fee no infcription on his monument, yet its filence proclaims more than all thofe of his brothers.

The fepulchre of the King and Queen has an altar near it, dedicated to the Holy Crofs: thofe of the Infantes, in like manner, have altars, which are feparated from each other, by arches made in the receffes of the wall oppofite the King's feet. Thefe altars are confecrated to their refpective tutelar Saints, and decorated with fmall altar-pieces, fuitable to the fituation, and with ancient paintings in good prefervation. The firft, which is next to the fepulchre of the holy Infante, is dedicated to the affumption of our bleffed Lady. Among the pictures is feen the portrait of this Prince, with his chains and the incidents of his misfortunes.

The fecond is confecrated to John the Baptift, and is applicable to the name and devotion of the Infante Don John.

In the third, the Infante *Don Anrique*, ordered the Infante Don Ferdinand to be painted; for he confidered him a Martyr, and as fuch rendered him devotion.

On the altar of the Infante *Don Pedro*, which is the fourth, is a painting of the Angel Saint Michael, whofe infignia he adopted for a device, as we obferved before.

The wall oppofite to the King's head is occupied by a large facrarium, in which, are preferved the holy utenfils, ufed in the celebration of the maffes daily offered up for thefe Princes, and diftinguifhed by their refpective devices and mottos.

That we may not omit any thing concerning the holy Infante, we find here among his wreaths, what was deficient in his fepulchre, namely, the motto which like the reft is in French, *Le bien me plait*, that is to fay, *I am pleafed with the good*. This he verified by his death and actions. But confidering his magnanimity, I can hardly believe that he ufed this motto, or permitted it to be applied to him whilft living; except that the branches in his emblem are of ivy, and

D

not of briar as has been afserted, which appears the more probable from their having no thorns. Then, this motto is proper, because there are two forts of ivy; one of which grows clinging round plants to the fummit, however lofty, without injuring them, and draws its fuccour from that which it intwines. The other is of a different quality, for it injures the trunk which it intwines. It appears by the motto, that he is only pleafed with the good fort, which is confiftent with a man, who from the cradle to the grave, was exemplary for his fteady fanctity of manners. It is alfo conformable to the purity of his foul, confidering, a peculiarity which is attributed to this plant; namely, if wine and water mixt, be thrown into any veffel made of its wood, the wine will difappear and be loft, and clear water only remain*.

Defcription of the Church and Monaftery externally.

THE Church has two entrances, viz. the principal and the tranfverfe. The porch of the principal entrance alone would require a volume to particularize the columns, figures, and variety of ornamental fculptures, with which it is decorated.

In the centre of the weft front, juft above the porch, is a window of fuch exquifite workmanfhip, that it is fcarce poffible to execute the like, with more accuracy, in wax or needle work, or in the overture of a guitar †. The laft comparifon is very applicable to its circular form, and to the minutenefs of its parts: the intervals correfpondent to thofe, which in the guitar emit the interior found, are here filled with ftained glafs, reprefenting in various paintings the arms and devices of the kingdom; together with fymbols and emblems relating to the King. As the perforations are numerous, because of the magnitude of the window, they admit a profufion of light, though fomewhat obfcured by the colours; but this obfcurity is amply compenfated by the beauty of the variegated rays.

It is furprifing, how fuch delicate work has ftood, fo many years, uninjured, in that lofty fituation. The durability and grandeur of the other windows of the Church are no lefs furprifing. In the body of the Church alone are 30 windows, fo large, that in a clear night the Church is almoft as luminous as an open fquare, notwithftanding the glafs is covered with colours.

It will not be amifs to defcribe the dimenfions of fome of thefe windows, which, we had taken by an Architect, in confirmation of what we affert. In the upper part of the Nave, there are 16 windows, that is to fay, eight at each fide, of 18 palms in height to the capitals of their columns, by 9 palms in breadth,

* Plin. b. xvi. c. 19.   † See the Weft Elevation of the Church.

OF THE ROYAL MONASTERY OF BATALHA.

with two mullions, of one palm each, in thickness, to secure the glass; which being deducted out of the breadth, leave 7 palms of aperture; these, multiplied by the height, make 126 square palms, the quantity of glass in each window.

The two Ailes have 12 windows; that is to say, four to the south, joining King John's chapel, and eight at the opposite side, each twenty-two palms high by seven and a half broad. And, because they are likewise divided by two mullions of one palm thick, like those of the Nave, each window consequently contains 121 square palms of clear aperture, and the same quantity of glass.

Of the same height and breadth with these, are two windows that accompany the principal door, one at each side, which added to the others make the number we mentioned of 30. Such a quantity of glass may be reckoned one of the most remarkable things in the edifice.

To the above number, are to be added two windows in the Transept; that over the transverse entrance alone is 42 palms high, by 14 broad, finished in a most curious manner with reticulated stone work, and the interstitial vacuities, filled up with stained glass*. These, together with the windows of the centre and collateral chapels, added to the fine window over the principal entrance, render the Church luminous, cheerful, and beautiful.

About the time the Church of Batalha commenced, was founded the famous temple of the see of Milan, called *Il Domo*, during the pontificate of Urban VI. The interior of this temple is altogether dull, and melancholy; contrary to that of Batalha, which is light and cheerful. The Milanese defend their artists, and attribute the effect to intention and sound judgment; alleging, that as a grave, solemn aspect, renders a man more respected, so, in like manner, a degree of gloominess makes a Church more venerable. This reasoning, however, does not convince me; for, allowing the argument to be true with regard to men, it does not appear that temples, which are the representations of heaven, the seat of eternal light, should have any resemblance to the abode of perpetual darkness. To return to our history: all these windows are so securely fixed, so transparent and beautiful in colour, that, notwithstanding they have already been exposed to the injuries of 200 years, they scarce exhibit any traces of decay.

The stone vaults of the Church are also covered with a roof of the same materials, composed of large square flags, about the thickness of the leaf of a strong table, and *rebated* at the edges †: thus, an everlasting

---

* See the Elevation of the *Transept Entrance*.   † In the *Transverse Section* of the Church, the covering of the roof he alludes to is represented.

roof is formed, that will bear to be walked over, swept, and cleansed of every dirt, contracted by the accumulation of years.

The top is enclosed on every side with a railing of stone work, over which are placed *fleurs de lis*, that serve as a distinct crowning. In order to see every part of this vast machine, there are two circular stair-cases, formed like screws, which lead to the roof of the building; one is in the transept wall, at the left-hand side on entering at the south door; the other adjoining the chapel of our Saviour; each consisting of 120 steps*, from the church floor to the summit of the edifice.

There is also another stair-case in the convent of very easy ascent, from the summit of which, we have a remarkably delightful and extensive prospect of a mountain of stone. Nor does it differ very much from other mountains, except, that here, the rocks are worked and polished by dint of art; whereas the others are shapeless, in a rude state of nature, with various inequalities, now sinking into deep valleys, then proudly aspiring to the clouds. In the same manner we see some inequalities in this structure: in one part a mountain appears, as in the body of the church, whilst other parts are in a valley, as the refectory, chapter-house, and cellar; lofty pyramids are also found here, emulating in height those of nature, but much superior in beauty.

These pyramids, which are three in number, are of very rare workmanship, and constructed in such a manner as to admit of access to the summit †, but not without danger, because of their great elevation: the one over the lanthorn of the Founder's Chapel is in the form of an octangular pyramid ‡, 50 palms in height, with a space of 4 palms on each side of the base, surrounded with a parapet of net work, and crowned with *fleurs de lis*. Taken altogether, it has a majestic and beautiful appearance. Another takes its rise over the Archives, between the cloyster and the sacristy, and is in height 63 palms §. The spire over the tower ||, which contains the bells and clock, is not less deserving of notice; it is similar to those we have just mentioned, and suitable to the rest of the fabrick.

---

* Each step is 9½ inches high, the whole height is consequently 95 feet.

† There is no other access to its summit than by means of the bunches placed on the angles of the pyramid or spire (figures 15 and 16 in the Plate of Ornaments, Mottos, &c.), to which I suppose he alludes. Sir Christopher Wren was of the same opinion. "The angles of pyramids in the Gothic architecture were usually enriched with the flower the Botanists call *Calendua*, which is a proper form to help workmen to ascend on the outside to amend any defects, without raising large scaffolds upon every light occasion. I have done the same, being of so great use, as well as agreeable ornament."

Parentalis, p. 302.

‡ The pyramid he alludes to was destroyed by the earthquake of 1745, as observed in the Preface. I have given a representation of it in the *South Elevation of the Mausoleum of King John the First*, from an ancient painting on one of the windows of the church.

§ This spire is 62 feet 2 inches high, as represented in the plate; which height I determined two ways, first, by a quadrant; secondly, by taking the diameter of the base of the pyramid, and the angle of inclination of one of its sides.

‖ That spire does not exist at present.

Of the Sacrifty, with the treafure of relicks, gold, filver, and veftments, which the Founder beftowed.

THE entrance into the facrifty is from S. Barbara's chapel, at the end of the crofs. It has nothing in its architecture remarkable for grandeur or compofition, but is well worth feeing on account of the treafure of facred relicks, gold, filver, and veftments of brocade, tiffue, and filk of every kind, that the King, with a true princely liberality, conferred on it: we fhall commence with the moft precious, which are the relicks.

The Emperor of Conftantinople, *Emanuel Paleologus*, being in the city of Paris in France, in the year of our Lord 1401, whither he came, for the purpofe of obtaining the joint affiftance and protection of the Chriftian Princes of the Weft, againft the power of the Ottoman Empire, which was overrunning Afia, and threatening Conftantinople and Europe: the King fent an Ambaffador to vifit him, which vifit the Emperor returned with a prefent of precious relicks, very eftimable for their quality, and for the certitude and credibility they derived from the authority of fo great a Prince. Thefe relicks were accompanied by a Certificate figned with the Emperor's own hand, and a feal of gold pendent. We fhall here give a tranflation of this Certificate, as it well deferves our notice.

" *Emanuel Paleologus*, Faithful in Chrift, Emperor of the Romans, and ever
" Auguft. Health to all to whom thefe imperial letters fhall come, through him,
" who is the falvation of all. Our merciful Saviour and Redeemer Jefus
" Chrift, who offered himfelf to God the Father, as a facrifice without ftain,
" at the altar of the holy crofs, left to all faithful Chriftians the marks of
" his fufferings as a remembrance of his miracles."

" Whereas we had in our city of Conftantinople, fome holy relicks of
" our Saviour, and of many Saints worthy of veneration, as handed down
" to us from the moft Serene Emperors our Fathers, in authentic hiftories
" and faithful chronicles; the faid relicks being preferved by them, as they
" are likewife by us, with all due care and veneration. And, whereas we
" happened to fojourn to thefe Weftern parts, on account of the perfe-
" cutions and oppreffions of the Turks, the cruel enemies of the moft facred
" name of Jefus Chrift, which they, with their forces, endeavour to extermi-
" nate from the face of the earth, efpecially in parts of Thrace: therefore, in
" order to feek aid and protection for the Chriftians of the Eaftern pro-
" vinces, who are oppreffed by the faid infidels, we brought with us part
" of the aforefaid relicks and fhrines; and being perfuaded, that the moft

E

"illuſtrious Prince, our couſin, Don John by the grace of God, King of Por-
" tugal, is worthy of all honour, and filled with zeal for the faith and
" Chriſtian religion, inſomuch, that his devotion for the Lord daily increaſes,
" we thought proper to give him part of the ſaid ſacred things, and ac-
" cordingly, we now give to the ſame moſt Serene Prince a ſmall croſs of
" gold, within which are ſome precious relicks of the Apoſtles, *St. Peter,*
" *St. Paul, St. George,* and *St. Bras*. In the middle of the ſaid croſs,
" is a particle of the ſponge, with which the gall and vinegar were given
" to our Saviour. In teſtimony of the truth of what is herein delivered,
" we have cauſed this paper to be written to the ſaid moſt Serene Prince,
" ſigned with our hand in Greek, and in red ink, according to the cuſtom
" of our Empire, and authenticated by our ſeal pendent of gold, bearing Greek
" letters. Dated in the City of Paris, the fifteenth day of June, 1401.
" We alſo give to the above King, a ſmall remnant of the garment of our
" Redeemer Jeſus Chriſt: it is of a colour reſembling the yew, of which tree it
" is made. Any woman afflicted with the hemorrhage, is immediately cured
" by touching its hem. This ſacred relick is incloſed in a ſhrine of cryſtal ſet
" in gold."

" *Emanuel Paleologus.*"

THESE holy relicks, accompanied with the certificate of the Emperor, the King received, and commanded to be depoſited in the Convent and Sacriſty. The ſeal is round; on one ſide, is a large Latin I, placed in the centre, with a medal, bearing the portrait of the Emperor, and an inſcription, *Emanuel, in Chriſt, Emperor Paleologus*. On the reverſe is an image of Chriſt, with another Latin I, and the words, *Jeſus Chriſt*. The Latin I, ſhews the title, in which, he eſteemed himſelf the Emperor of the Romans.

Such are the Relicks; we ſhall next point out the plate and other articles preſented by the King, viz. Fifteen ſtatues of caſt ſilver, very coſtly and beautiful, repreſenting the like number of Saints of his devotion. Twenty-eight chalices, moſt of which are gilt. Fourteen pair of cruets. Five fonts with their ſprinklers. Eight cenſers and ſix boats. Nine ſmall croſſes to accompany them. Nine portable croſſes for the ſervice of the altars. Four great croſſes, three of which are for proceſſions, and one to ſtand on the high altar. Two lofty candleſticks gilt, and twelve inferior ones. Six torch ſtands, of which two are gilt: it is recorded theſe two weighed ninety-one marks. Seven lamps of immenſe ſize and weight. One lantern. Five pixes. Five paxes. Two ewers, with their water plates for waſhing the hands. And two ſmall bells.

This plate, as may be well conceived, weighs more than 1200 marks, and is very valuable for its workmanſhip, and on account of a great part of it being gilt. Reduced to our standard, it weighs upwards of 600lbs. a magnificent and royal donation for the service of the house of God, and that, at a time, when we possessed neither East nor West Indies.

The furniture that he commanded to be made, for the service and celebration of the mass, and for decorating the altars, consists of eleven very rich brocades, with their copes, altar and pulpit cloths, the greatest part of which are ornamented with fancy lace work, or embroidered with gold in a very rich manner. Thirty-two vestments of costly silk, together with many particular vestments of brocades, tissues, and silks, for the ordinary service of each day. Besides several large cloths of gold, of brocade, and of velvet, and different curtains of silk, to ornament the church, altars, and to cover the sepulchres, at the celebration of the anniversaries.

A great deal of the plate being superfluous, there was sold as much of it, as weighed 811 marks. Of the vestments, there were sold, altogether, four of the richest, and another was melted down, which was covered all over with silver foil, in such a manner, that none of the silk under it was perceptible, so that its weight rendered it fitter for ornament than use. The committee appointed for disposing of the above, were composed, not of the Fathers, but of people called in, who judged it necessary to raise a sum of money to support and carry on the work of the Convent, which was at a stand. Having obtained licence from the Penitentiary's Court at Rome, addressed to the Bishops of *Lamego*, *S. Thome* and *Targo*, in the fourth year of the Pontificate of *Pope Paul* III. they commanded the overseers and melters to purchase amalgamating implements; very necessary things for a Convent that possesses but little revenue.

Description of the Chapter House*.

FROM the Sacristy you enter the Chapter House. This room is so constructed, that there can be nothing more wonderful, in so much, that it comprehends the utmost degree of architectural skill. Its form is a square, each side of which measures 85 palms, and is covered with a vault of hewn stone without column, prop, or any thing to support it, but external buttresses, such as are in the side of the church.

It is recorded, that in constructing this vault, it fell twice in striking the centres, with great injury to the workmen. But the King, desirous, at all events, to have a room without the defect of a central support, promised to reward the Architect, if he could accomplish it. At this, he was animated in such a

* Annexed is the Section of this magnificent room.

manner, that he began it again, as if confident of fuccefs. The King, however, would not hazard any more the lives of his workmen in ftriking the centres: therefore he ordered, from the different prifons of the kingdom, fuch men, as were fentenced to capital punifhments, in order that, if the like difafter happened a third time, none fhould fuffer, but thofe, who had already forfeited their lives to the offended laws of their country.

In the middle of this room, is a large quadrangular eftrade of wood, with fteps on every fide: over this eftrade, are two tombs, covered with rich cloths: in one are the remains of King Alfonfo V. grandfon of the founder; in the other is depofited Prince Alfonfo, fon of King John the Second, who unfortunately loft his life at *Santerem*, by a fall from his horfe as he was riding along the banks of the Tagus.

Royal Cloyfter.

FROM the Chapter Houfe we pafs to the Royal Cloyfter: the door of the one is oppofite to that of the other; its form is a fquare; each fide meafuring 250 palms, 30 of which are covered at each fide with ftone vaults. The vaults of this Cloyfter exhibit a number of magnificent Gothic arches, filled up, from the fpring to the fummit, with beautiful tracery work: it is evident from their excellence, that the Artifts who executed them, were not inferior to thofe who formed the frontifpiece of the edifice, nor the defigners of the former, lefs ingenious, than thofe of the latter.

The infcribed quadrangle of the Cloyfter is diftributed into walks, bordered with large hewn ftones, and the inclofed fpaces are planted with a diverfity of fhrubs and flowers. In the midft is a large ciftern of water; and in one of the angles, a lofty fountain, with water jetting out of different bafons, into a large circular refervoir. The fountain is very ufeful in this fituation, becaufe, the refectory door is contiguous to it, fo that thofe who enter, may wafh their hands, and gratify their fight, whilft waiting the fignal of the dinner bell: for this purpofe, the wall next the refectory door is furnifhed with feats and wainfcoted backs, for the accommodation of the Fathers.

Refectory.

THE Refectory door is fituate in the N. W. angle of the Royal Cloyfter, the weft fide of which, it nearly occupies. This room, from its magnitude and excellent workmanfhip, may be confidered of a piece with the

OF THE ROYAL MONASTERY OF BATALHA.

rest of the fabrick, being in length 133 palms and 44 in breadth*, which is nearly equal to one third of its length: it is lofty and well lighted: the whole is covered with a stone vault similar to those we have already described.

The other inferior offices of the convent in general, such as the store rooms and cellars, are of dimensions suitable to their use, and sufficiently large to contain the various fruits for the sustenance of the society, who are numerous on account of its being a place of continual study. The wine cellar alone is 160 palms long by 43 wide, covered with a vault.

North of the wine cellar there is a cloyster of about half the area of the Royal cloyster, of very good workmanship. On viewing the outside of the convent it appears like a fine village, or many convents together, because of the dormitories, *hospitia*, infirmaries, library, and the novices' preceptory. The last is extensive enough for a good convent, from the dimensions of its corridors, the number of its cells, the distribution of its oratory, together with its orchard and garden.

Through the dormitory of the convent runs an extensive corridor, covered with a common roof, and lined with boards, the better to preserve the Fathers' health. At the north end is an open terrace, which commands a pleasant prospect of orchards, and a large vineyard, refreshed by the constant course of a fine river. Here also are seen several deep ponds, that at times afford amusement to the recluse and studious Fathers by fishing with cane rods and nets. In this corridor are the infirmary and hospitium, with more than 60 cells, exclusively of those of the novices which amount to 24. The lay-brothers' apartments both externally and internally, the various entrances, the passages to the different offices, the several involutions, notwithstanding their number, all have the appearance of a Royal edifice. Order and convenience, cleanliness and good attendance, are observed in every part; and, to complete the whole, a large subterraneous stream of water runs across the edifice, which carries off all the impurities of the convent.

* 32 palms is the breadth mentioned in the original, which is a mistake.

Defcription of the Maufoleum of King Emanuel.

WITH the foregoing humble and unadorned defcription, we have endeavoured to exhibit the magnificence of the edifice which is completed; but there is another lefs ancient not yet finifhed, which, if perfected, would crown the fplendor of this monaftery.

At the back of the chapel of S. Barbara [*], there is a door, with a crofs over it in demi-relief, like thofe worn by the Knights of the order of Chrift: this crofs ftands between two armillary fpheres, which guard a fhield bearing three characters, viz. the letters Ey within a large C[†].

One would fuppofe that the author of the work intended to excite the attention of the curious who came thither, and was refolved that his enigmas fhould coft them more trouble to decypher, than an Egyptian hieroglyphick, or a Sibylline oracle. Indeed, it would be much eafier to form a judgment of fubjects of the latter kind, becaufe, with the affiftance of words and allufive figures we may draw fome fatisfactory conclufion. But from a few infulated letters it is difficult to afcertain the precife meaning of the author; and the more fo, as they are fubject to receive, camelion like, the colours we wifh to impart to them, or, like virgin wax, fufceptible of every voluntary impreffion.

This door of emblems and myfterious cyphers leads to a loggia directly behind the high altar of the church [‡], in the middle of which is a beautiful door-way [§], compofed of fmall ornamented columns, which are formed into arches over-head: they have no capitals, nor any other feparation whatever. The entrance, which is very wide, has feven columns on each fide, and commences, from the extreme ones at the outfide, to contract as it recedes till it attains the magnitude of a moderate aperture.

The dimenfions and ornaments of the feven columns are various, but all fculptured with fuch exquifite delicacy, fo beautiful in form, and perfect in execution, that one would fuppofe it impoffible to form the like of the moft pliant wood ||. The work, therefore, muft have been attended with great expence, confidering the time and labour it took to carve and polifh every ftone,

---

[*] See the letter Q in the *General Plan*.
[†] Vide Fig. 2 in the annexed Plate.
[‡] P in the *General Plan*.
[§] See the Entrance to the Maufoleum of Emanuel the Great.

[‡] In the annexed Plate we have given a reprefentation of thefe columns, with the other ornaments and hieroglyphicks appertaining to this Maufoleum.

and the number that muſt have been broken under the chiſel, on account of the extreme delicacy of the parts.

In four of the interſpaces of the above columns, there are words at certain intervals, the characters of which are ſimilar in form to thoſe we mentioned before: they are as follow, *Tanyas erey* *. I have been the more particular in noticing theſe characters, becauſe in paſſing from this monaſtery formerly, in our journey to *Entre Douro e Minho*, we met at *Vianna* an edition of a book entitled, The Life of the Holy Archbiſhop of *Braga, D. F. Bartolomeu dos Martyres*. The Fathers of that place read ſome words in it thus. *Tangas e Rey*, making a *g* of the *y*, in *Tanyas*, and dividing *erey* into two words, which evidently is contrary to the form of the ſculpture, as I then ſhewed them. The frequent repetition of theſe words made me ſolicitous about inquiring after their meaning among the old Fathers, hoping, that by tradition from their predeceſſors, they might have obtained a knowledge of the ſenſe that was annexed to them at the time they were ſculptured. And this did not appear at all improbable, becauſe they are not more ancient than the days of King Emanuel. Of the language it is not eaſy to form any opinion, as it is neither Latin nor any of the modern European tongues. But we could find no perſon to ſatisfy us with any well founded argument reſpecting thoſe, or the other cyphers at the entrance of the Loggia.

As it is the duty of an author to deliver his opinion on doubtful paſſages of hiſtory, it will not be deemed preſumptuous in me, to endeavour to untie, or cut with a ſhort diſcourſe, this which is not a Gordian knot, though it belongs only to Alexanders to touch thoſe knots which are tied by Kings.

As I found in theſe words no affinity to the language of the country, I had recourſe to foreign languages; and upon ſhewing the words to a perſon of great erudition, we concluded that they were Greek; *Tanyas* being the accuſative of the Greek word *Tanya*, which ſignifies *Region*, and *Erey* the imperative of the word *Ereō*, which ſignifies either *Seek thou*, or *Inquire*, or *Diſcover*,—ſpeaking as it were in the name of the Lord, from the Temple which King Emanuel built, ſaying, *Go thou and explore new regions and new climates*; thus animating him, not to deſiſt from his glorious deſigns. And this ſignification is conformable to the enterpriſe this Prince had actually in contemplation at the time, namely, the diſcovery of India †. It alſo agrees with

---

* Vide Fig. 9. of the Plate of columns, ornaments, and hieroglyphicks.
† During the reign of John the Second, the Cape of Good Hope was diſcovered by *Bartholomew Diaz*. Soon as Emanuel his ſucceſſor aſcended the throne, he aſſembled his council to deliberate on the expediency of proſecuting the diſcovery of India, which, after much counſel, was reſolved in the affirmative. Vaſco de Gama was appointed to take command of the fleet ſent upon that expedition. He ſailed from Liſbon the 8th of July, 1497, with three ſmall ſhips and 160 men, and arrived in India the year following.

the device of his myſterious ſpheres, which were adopted by him for another end, and were prophetic of his conqueſts over a great part of the globe.

But the cypher of the firſt mentioned door gives us more trouble, as it conſiſts of letters which form no definite word. The firſt difficulty that occurs, is to aſcertain the language to which thoſe characters belong. I take them to be Greek, like thoſe we mentioned before; and, being ſurrounded by ſpheres and the croſs of the order of Chriſt, I am induced to conclude there is unqueſtionably ſome myſtery inveloped in them.

It appears, indeed, as if the founder of the work intended that we ſhould have here, an emblem in imitation of the celebrated ancient temple of Delphos in Greece. For, we read, that over the door of that temple, there was a cypher almoſt ſimilar, and at the portal an inſcription addreſſed to thoſe who entered. The inſcription was this, γνωθι σεαυτον; that is to ſay, *Know thyſelf*. The cypher was ΕΙ, which ſignifies, *Thou art*. This cypher ſo puzzled the ancient Sages, that Plutarch compoſed a book upon it*, wherein, after many arguments, he concludes, that by the word ΕΙ is meant *One Eternal God*. His words are as follow:

Deus enim eſt, et eſt nulla ratione temporis, ſed æternitatis immobilis, tempore et inclinatione carentis, in qua nihil prius eſt, nihil poſterius, nihil futurum, nihil præteritum, nihil antiquius, nihil recentius: ſed unus cum ſit, unicus nunc ſempiternam implet durationem.

And lower down,

Non enim multa ſunt Numina, ſed unum.

Which nearly ſignifies, " that it is only of God one can and ought to ſay that He is; and this Being is without any dependence or meaſure of time, a permanent and immutable eternity, an eternity without time or change, inſomuch that there is nothing in it firſt or laſt, nothing paſt or future, nothing more ancient or more modern: as He is but one, with one only His preſence now fulfills, accompliſhes, and declares His perpetual eternity without end."----" Becauſe in truth there is but one God."----Thus far Plutarch.

The learned Father *Franciſco de Mendoça*, of the Society of Jeſus, favours this meaning in his firſt volume upon the Book of Kings; and alſo Euſebius in his Evangelical Preparations †. It is a doctrine ſo conformable to what we find in holy writ, that one would think the Gentile Hiſtorian took it from that part where we read *Ego ſum qui ſum*, QUI EST *miſit me ad vos:*

---

* Plutarch. lib. de ει, apud Delph.   † Euſeb. lib. ii. c. 7.

I AM THAT I AM. I AM hath sent me unto you*. This author finally concludes that the cypher EI is an admonition to us that we are obliged to fear, love, respect and adore a God who exists through all eternity. The words which close the treatise say as much:

Hoc enim pronunciatum est, ut nos percellat & ad venerationem Numinis, utpote quod sit semper, excitet.

Hence I have no doubt but the same thing is represented to us in the emblem of which we speak, and that it is the answer of the King to the Lord, who commands him to undertake the discovery of new seas and lands, saying, as it were, *I know, O Lord! that nothing but thee is worthy to be sought for. These seas and lands, even though joined by many worlds, have their limits and duration, but thou alone art eternal, immortal, infinite.*

Nor does the letter c oppose this meaning, for it serves only to support the other two, or to point out to us, in the eternal essence, the sacred mystery of the Trinity of the Divine persons, unknown to the Pagan historian. This we shall attempt to shew in two ways; first, because it is the third letter of our alphabet, the same mode of reasoning by which Plutarch proves that the cypher EI was the number five. In the second place, because with the other two letters, it makes the number three, as it surrounds the E to which the *y* is joined.

The third letter is selected from the Latin alphabet in preference to the Greek, and joined with the other two Greek letters. Though at first this may appear an impropriety, we may rest assured it was dictated by judgement: it is a testimony of the catholic truth, expressly adopted by one who was a son of the Latin church.

To conclude, I have only to observe to those who have a knowledge of the Greek, not to be surprised if they find the Greek *y*, where the Latin *i*, or the Greek *iota* should be: we ought to pardon such inadvertence in the illiterate workmen who carved them, since in our language and mode of writing both letters have the same power and signification.

Having passed through the door of the Loggia, we behold a large unfinished edifice which forms a perfect octagon †. In this edifice are seven chapels similar in design and execution; their arches, tracery work, and various ornaments are finished in the most exquisite manner. Indeed it is much to be lamented that such a structure should remain incomplete, considering the ex-

---

* Exodus, c. iii. v. 14.   † See the letter R, in the *General Plan*.

pence of finishing it would not be very great in comparison to what has been already done, as the walls of the entire are raised to that height, whence, according to the rules of art, the large vault should spring to cover the fabrick, and make that which is now an open area an inclosed chapel. Notwithstanding this fabrick has been so many years exposed to the weather, it scarcely discovers any marks of decay: an evident proof of the goodness of the workmanship and durability of the materials.

The purpose for which this pile was raised in so magnificent a style is very evident. The remains of all the Princes after King John the First and his Sons, are disposed about the convent, in a manner that scarcely deserves the name of interment. It was but reasonable, therefore, that some pious heir in the tranquil enjoyment of peace, should collect their remains, and give them a friendly and suitable abode.

Now there are various opinions respecting who it was that conceived the idea of the edifice and put the first hand to it. But it is not to be doubted that the greatest part of it was done by King Emanuel, or at least with his permission, and during his reign, for we have positive proofs to verify it. In the first place, we see in the chapel opposite to the entrance* (the most conspicuous of the seven) two spheres similar to the above mentioned over the first door, the well known constant devices of King Emanuel. In the second place, we read upon shields, in the pendent orbs of the vault of his chapel, the words *Tanyas erey*. Hence we may conclude, that this inscription, so often repeated in the beautiful portal, together with the emblem of three letters over the first door, evidently belonged to the same King, for both are accompanied by the spheres. The cross of the Order of Christ seen over the first entrance is not averse from this argument, for he was Master of that Order before he was King, and afterwards united it for ever to the crown. But all our doubts respecting this matter are removed by a Latin Inscription which is over a door in the Loggia contiguous to the entrance: it is this, *Perfectum est opus anno* 1509, meaning that it was brought to that state of perfection in such a year, being the epoch in which he attained a good age and a glorious reign, enjoying the victories and treasures of India †.

There are some however who think the founder of this work was Queen *Laynor*, sister to King Emanuel, intending it as a depository for the remains of her husband, King John the Second, and her son, Prince *Alfonso*, neither of whom have proper sepulchres in the convent: having possessed a large revenue, and was a Princess of great piety. King Emanuel, her brother, was much at-

---

* T, in the *General Plan*.
† There is a stronger proof than any he has yet adduced that this chapel was built in the time of King Emanuel, that is, the letters M. R. (*Manuel Rex*) often repeated in demi-relief upon the architraves of all the upper windows of the Chapel. (Vide Fig. 17. in the Plate of Columns, Ornaments, &c.)

tached to her, as well from kindred, as on account of particular obligations he owed her, for the decided part she took in promoting his acceſſion to the throne, from which King John his predeceſſor was manifeſtly averſe. Thoſe who hold this opinion add, that ſhe reſerved the beſt chapel (which is that oppoſite to the entrance) for King Emanuel, and diſtinguiſhed it by his mottos and devices, which are very different from the ſober modeſt ſtyle of King John the Firſt. The adjoining chapel ſhe intended for herſelf and her conſort. In this chapel we behold her ſapient device, which is the Pelican in the act of piercing its breaſt. But time, the ſovereign arbitrator of all human projects, has evinced the inefficacy of her deſigns, and ſhe has loſt the merit of the work by leaving unfiniſhed and expoſed to the weather an edifice worthy of being tranſmitted down to the lateſt poſterity.

I ſhould not forget to mention the ſentiments of a perſon of profound judgment reſpecting this building, and the entire monaſtery; I mean the grand maſter, *Cardinal Vincent Juſtiniano*, our General, who has ſeen and eſtimated the merits of moſt of the principal edifices of Europe, and whoſe teſtimony we can the leſs doubt, becauſe he was a ſtranger, and a perſon of the moſt religious veracity and candor. When this Father came hither, he made obſervations upon every thing worthy of notice in the Kingdom. Upon ſeeing this convent he exclaimed with admiration—*Videmus alterum Salomonis Templum.*

Of the Offerings the Founder bequeathed to the Convent.

ON the anniverſaries of the King and his Sons, offerings are allowed to the convent, conſiſting of a certain quantity of wheat, wine, and wax. And as the order of this convent originally abſtained from fleſh-meat, the pious King wiſhed likewiſe to add an offering of ſome dozens of dried whitings, of a large and wholeſome ſpecies. Theſe are of great ſervice to the community, and eaſily obtained, as the ſea-ports near the convent produce fiſh in great abundance. As the anniverſaries are many, the offerings are princely, they amount to fifty-two *moyas* and a half of wheat*, forty-three pipes of wine, twenty-four *arrobas* of wax †, and two hundred and fifteen dozen of fiſh. Theſe offerings, reduced to money, the King commanded to be paid quarterly out of his revenue by the receiver of the diſtrict of Leiria. Since the prices of theſe articles have now increaſed, they amount to a conſiderable charity, and are at preſent the principal ſuſtenance of the Fathers.

* A Moya is about 11½ buſhels, as we obſerved before.   † An Arroba is a weight of 32 ℔. in Portugal.

Of the Ordinations the King commanded to be observed in this Convent by his Will.

"HAVING considered the manner the Fathers of S. Dominick usually live in their communities, we command that the following ordinance be observed in the completion of the said Monastery, and for the good support and maintenance of the said Fathers. For this purpose we command and beseech the Infante *Duarte* my son, or whoever may hereafter be King and Lord of these realms, that he fulfill and observe every thing after the manner by us prescribed. First, we command that the said Monastery be finished with a cloyster, dormitories, and all other necessary apartments. That they be of such ample dimensions, as shall be found necessary, and, that the expences of the same be defrayed out of the rents of *Leiria* and its district. That there be maintained and continued the same number of Friars which are in it at this time, and in the manner in which they are supported at present. That they observe the same manner of praying, and saying their masses, responses, and making funeral processions for my soul, and that of the Queen my wife——the Lord rest her soul! That they offer for the benefit of my soul, after interment, such masses and prayers as the said Infante, or whoever may hereafter be King of these realms, shall ordain in the Monastery. And, that the number of thirty Fathers be kept in it, and supported in the undermentioned manner, and from henceforth observe the regulations by us ordained. And when the said Monastery is complete in all the foresaid necessary works, with the rents and taxes of Leiria and its district, that there be drawn thence such sums as shall be found requisite for the maintenance of the said Fathers, and for the purchasing as many estates and holdings, as may decently support, and provide eating, drinking, clothing and shoes, for the said thirty Friars of the Dominican order, to wit, twenty of sacred orders, ten Novices and Lay-Brothers, besides a certain number of servants, and likewise a baker, cook, muleteer, laundress, shoemaker, and such like persons as may be thought necessary; and these thirty Fathers we command to live continually in the said Monastery."

Of the Regulations the King ordained for the Government and Preservation of the Edifice.

THE Pope, through the entreaty of the King, granted particular graces and indulgences to the Novices who take the habit in this Convent, or receive their education, or die in it. And as to bodily affairs, he did not forget to provide the Fathers with essentials. He ordained that a physician should reside continually in a neighbouring place, whence he could hasten to the indisposed. To this end he was obligated, not only by a competent salary, but likewise by certain honours, and the privilege of a phyſician to the Royal Houſchold. This he left eſtabliſhed as firm as law could enact. Nor did his prudence forget the neceſſity which all great edifices have for continual reparation; therefore he ordered an agent, with the title of Surveyor of the Works, to reſide in the vicinity, to whom a number of workmen of different trades were allowed, as often as there was neceſſity to build or repair any part, and theſe he honoured with certain exemptions and privileges. And, that there might be no faults nor delays in accompliſhing every thing that was wanting, he commanded that they ſhould be numerous, to wit, 125 ſtone-cutters, 56 quarry-men, 20 carters, 10 labourers, 1 ſmith, and only 2 carpenters; as we obſerved from the beginning, that there is no timber nor carpentry in the ſtructure, except the doors, all the reſt is ſtone and glaſs. For this claſs of people and their attendants, the honour of the privileges granted to them, was ſufficient to make them always ready without any other inducement, becauſe theſe privileges were always highly eſteemed; but when actually employed, they were paid their ordinary wages beſides.

Since the number and magnitude of the ſtained glaſs windows form a principal part of the beauty of this church; and as a thing ſo brittle is often in need of repair, the King aſſigned a particular ſum to a glazier to keep them conſtantly in order; in purſuance of which, he was bound to replace at his own coſt, whatever was damaged to the ſize of one palm, and all above that dimenſion was to be paid for in proportion, from the fund for ſimilar expences.

H

THE HISTORY AND DESCRIPTION

THE
EPITAPH OF KING JOHN I.

WE can never sufficiently acknowledge our obligations to this Prince, for giving us a place of abode, and choosing us for his perpetual chaplains. In testimony of our reverence for his venerable remains, which he ordered to be deposited amongst us, we subjoin the following memorial of his glorious achievements, word for word as engraved on his sepulchre by order of his son King Edward:

IN nomine Domini, sereniſſimus, et ſemper invictus princeps, ac victorioſiſſimus et magnificus reſplendens virtutibus, Dominus Joannes Regnorum Portugaliæ Decimus, Algarbii Sextus Rex, et poſt generale Hiſpaniæ vaſtamen primus ex Chriſtianis farnoſæ civitatis Septæ in Africa potentiſſimus Dominus, præſenti tumulo extat ſepultus. Excellentiſſimus iſte Rex, nobiliſſimæ ac fideliſſimæ Civitatis Ulixbonæ ortus Anno Dom. 1358, extitit per Sereniſſimum Dominum Petrum ſuum genitorem militaribus in ætate quinquennij ibidem decoratus inſignijs: et ſuſcipiens, poſt deceſſum Regis Ferdinandi fratris ſui, ipſius Lixbonenſis urbis et aliarum complurium munitionum, quæ ſe illi ſubdiderunt, gubernamen, obſeſſam perſonaliter per Regem Caſtellæ novem menſibus Ulixbonam mari grandiſſimā claſſe, et per terram ingenti vallatam exercitu, et plurimis Portugalenſium Regis Caſtellæ potentiam roborantibus circumſeptam, adverſus feras et multiplices impugnationes ipſam Ulixbonenſem civitatem ſtrenuiſſimè defenſavit.

Deinde nobilis civitatis Colimbricæ Anno Domini 1385, jocundiſſimè ſublimatus in Regem, per ſe et per ſuos bellicos proceres miranda exercuit guerrarum certamina: et pluries adverſantium dominia et terras intrando glorioſiſſimus triumphavit: et præcipuam, et regiam circa iſtud Monaſterium victoriam eſt adeptus: ubi Regem Caſtellæ Dominum Joannem, ſuorum maximo firmatum robore nativorum, et plurium Portugalenſium et aliorum extraneorum fultum ſubſidiis, iſte invictiſſimus Rex, virtute Dei Omnipotentis, potentiſſimè debellavit: et quamplures iſtius Regni munitiones et caſtra jam ſub hoſtium redacta poteſtate, viribus recuperavit armorum, uſque in ſuæ vitæ terminum virtuoſiſſimè protegendo. Et Deo recognoſcens, Glorioſiſſimæque Virgini Mariæ, Dominæ noſtræ, potiſſimam victoriam, quam in vigiliā Aſſumptionis obtinuit in menſe Auguſti, hoc Monaſterium in eorum laudem ædificari mandavit, præ cæteris Hiſpaniæ ſingularius et decentius. Et ſoli Deo optans honorem et gloriam exhiberi, et tantùm ipſi aut propter eum majoritatem fore cognoſcendam deſcriptionem, quæ ſuorum prædeceſſorum temporibus in publicis ſcripturis ſub Ærā Cæſaris notabatur, decrevit ſub Anno Domini noſtri Jeſu fore de cætero annotandam. Hoc actum eſt Ærā Cæſaris M.CCCC.LX. et Anno Domini 1422, tempore aliter defluendo.

Iſte feliciſſimus Rex, non minus reperiens quæ ſuſceperat regna illicitis ſubjecta moribus, quam ſævis hoſtibus, ipſa expurgavit cum diligentiā ſalutari, et propriis actibus virtuoſis uſitata facinora extirpando, pullulare fecit in his Regnis probitates honeſtas: et ſollicitus ad pacem cum Chriſtianis amplectendam, eandem ante proprium deceſſum pro ſe ſuiſque ſucceſſoribus obtinuit perpetuam. Et ſucceſſus fidei fervore, iſte Chriſtianiſſimus Rex, comitante eundem Sereniſſimo Infante Domino Eduardo ſuo filio et hærede, et Infante Domino Petro, et Infante Domino Henriquo, et Domino Alfonſo Comite de Barcellos, præfati Regis filiis, et ingenti ſuorum naturalium impavidā ſociatus potentiā, cum maximā claſſe plus quam ducentis viginti aggregatā navigiis, quorum pars numeroſior maiores naves et grandiores extitere triremes, in Africam transfretavit, et die primā quā telluri Afrorum impreſſit veſtigia, nobiliſſimam et munitiſſimam civitatem Septam oppugnando in ſuam poteſtatem redegit mirificè, et poſtmodo eidem urbi plus quam centum mille (ut aſſeritur) Agarenorum ultramarinis, et Granatæ pugnatoribus obſeſſæ idem glorioſiſſimus Rex per ſuos illuſtres genitos, Infantem Dominum Henricum, et Infantem Dominum Joannem, et Dominum Alfonſum Comitem de Barcellos, et alios Dominos, et Generoſos ſuccurſum niſit: qui fugantes de obſidione Agarenos quamplurimos in ore gladii trucidando; ipſorum claſſe ſubmerſione, incendio, et captura conquaſſatā; prædictam liberavit civitatem Septam: quam decem et octo annis minus octo diebus Anno Domini 1433, in menſe Auguſti, vigiliā Aſſumptionis Sanctiſſimæ Mariæ Virginis terminatis adverſus bellicos Agarenorum multiplicatos, inſultus validiſſimè præſultavit.

Menſe autem et vigiliā prædictis, iſte glorioſiſſimus Rex in civitate Ulixbonæ, aſſiſtentibus ſuis filiis et aliis quamplurimis generoſis, vitam feliciter complevit mortalem, relinquens notabilem urbem Septam ſub poteſtate altiſſimi potentiſſimique Domini Eduardi filii eius, qui paternos actus viriliter imitando, eandem in fide Jeſu Chriſti nititur proſperè gubernare. Iſte autem excellentiſſimus et virtuoſiſſimus Rex Dominus Eduardus tranſtulit honorantiſſimè corpus Chriſtianiſſimi Regis patris ſui, aſſiſtentibus eidem ſuis germanis, Infante Domino Petro Duce Colimbriæ, et Montis Majoris Domino; Infante D. Henrico Duce de Viſeo, et Domino Covillianæ, et Gubernatore Magiſtratūs Chriſti; Infante Domino Joanne Comiteſtabili Portugaliæ et Gubernatore Magiſtratūs Sancti Jacobi; et Infante Domino Ferdinando, et Domino Alfonſo, Comite de Barcellos, filiis præfati Regis Domini Joannis, qui tempore ſui obitūs alios non habebat, præter duas filias, quarum una erat Domina Infans Eliſabeth,

OF THE ROYAL MONASTERY OF BATALHA.

Ducissa Burgundiæ et Comitissa Flandriæ, et aliorum Ducatûum et Comitatûum : et alia Domina Beatrix Comitissa Hontinto, et Arondel, quæ in suis terris permanebant. Habebat autem Dominus Joannes nepotes qui Dominicæ translationi affuerunt, Dominum Alfonsum Comitem de Ourem, et Dominum Ferdinandum Comitem de Arrayolos, filios Comitis de Barcellos : et habebat nepotem Dominum Infantem Alfonsum primogenitum Domini Eduardi, et alios nepotes, et pronepotes, qui annumerati cum filiis erant viginti, tempore quo de præsenti sæculo migravit ad Dominum.

Affuerunt etiam hujus translationis celebritati omnes qui tunc in Cathedralibus Ecclesiis istorum Regnorum Prælati erant, et alii complures, cum multitudine Clericorum et Religiosorum copiosâ: et Domini et Generosi hujus patriæ, civitatum etiam et munitionum Procuratores extitere præsentes. Fuit autem venerandissimè delatum Regium Corpus ejus ad istud Monasterium trigesimâ die Novembris Anno Domini supradicto, et in Capellâ majori cum excellentissimâ, et honestissimâ, et Christianissimâ Dominâ Philippâ ejus unicâ uxore, prædictorum Regis Eduardi et Infantum, et Ducissâ illustrissimâ genitrice. Anno vero sequenti, die decimâ quartâ mensis Augusti, fuere per Regem Eduardum, et Infantes et Comites prælibata corpora prædictorum Regis, et Reginæ Philippæ cum honore mirifico ad hanc Capellam delata, quam ædificari pro sua sepulturâ imperavit, et huic deductioni extitere præsentes altissima et excellentissima princeps Domina Leonor horum Regnorum Regina, et Infans Domina Elisabeth Ducissa Colimbriæ, et Infans Domina Elisabeth uxor Infantis Domini Joannis, et præcipua pars Dominorum et Generosorum istius terræ, qui interfuerunt sepulturis prædictorum Dominorum Regis et Reginæ, quibus Deus suâ miseratione et pietate largiri dignetur sine fine fœlicitatem. Amen.

THE

EPITAPH OF QUEEN PHILIPPA.

IN gratitude to the memory of this most virtuous and excellent Queen, the consort of so great a King and the mother of so many illustrious Princes, we give the following Epitaph, which was dictated by her Royal Husband, and engraved on her tomb by command of King Edward her son:

HÆC felicissima Regina à puellari ætate, usque in suæ terminum vitæ, fuit Deo devotissima : et divinis officiis ecclesiasticè confuetis tam diligenter intenta, quod clerici et devoti erant religiosè per eandem sæpius eruditi : in oratione autem tam continua, quod demptis temporibus gubernationi vitæ necessariis, contemplationi aut lectioni, seu devotæ orationi totum residuum applicabat. Plurimum vero fidelissimè dilexit proprium virum : et moralissimè proprios filios castigando virtuosissima doctrinavit : et bona temporalia circa Ecclesias et Monasteria distribuendo pauperibus plurima erogabat ; generosis Domicellis maritandis manus liberalissimas porrigebat. Erat enim integra populi amatrix et pacis plena desideratrix, et efficax adjutrix ad pacem habendam cum Christicolis universis, et libenter assentiens in devastationem infidelium pro Dei injuriâ vindicandâ : et tantum prona etiam ad indulgentiam, quod nunquam accepit de sibi errantibus, nec consensit vindictam fieri aliqualem. Virtuosissima ista Domina extitit fœminis maritatis bene vivendi regulare exemplar, Domicellis directio, et totius honestatis occasio ; cunctisque suis subjectis fuit curialis urbanitatis moderatissima doctrix. In his autem et aliis quamplurimis perseverando virtutibus, quarum plurimitatem hujus lapidis humilitas nequiret ullatenùs præsentare, dictim et continuè pervenit ad istius vivendæ mortalis limitem ordinatum ; et fuit ejus vita fuit optima et valde sacra, sic mors extitit preciosa in conspectu Domini, et nimium gloriosâ : et, receptis laudabiliter omnibus Ecclesiasticis sacramentis, proprios filios benedixit, commendans eisdem quæ intendebat fore ad divinum obsequium et honorem et profectum istorum Regnorum, et quæ in eis sperabat causitura crementum indubiè : virtuosissimè, taliterque hujus mundi labores finaliter adimplevit, quod præsentes, qui relata audierunt, firmam suæ salvationis spem retinent singularem. Obiit autem decimâ octavâ die Julii Anno Domini 1415, et in Monasterio de Odivellas ante chorum Monialium decimâ nonâ die mensis ejusdem extitit sepulta : et anno sequenti, mensis Octobris die nonâ fuit prætiosum corpus ejus desepultum, integrum inventum et suaviter odoriferum, et per victoriosissimum Regem Dominum Joannem ejus conjugem, et per Serenissimos Infantes, scilicet, Dominum Eduardum suum primogenitum, et Dominum Petrum Colimbriæ Ducem, et Dominum Henricum Ducem Viscensem, et Dominum Joannem, et Dominum Fernandum, et Infantem Dominam Elisabeth, ipsius gloriosissimi Regis et felicissimæ Reginæ filios ; sociante Prælatorum, et Clericorum et Religiosorum copiâ numerosâ, et Dominis et generosis Dominabus, et Domicellis quamplurimis comitantibus, fuit corpus dictæ Reginæ honorandissimè translatum ad istud Monasterium de Victoriâ, et tumulatum in Capellâ majori et principaliori, die mensis Octobris decimâ quintâ Anno Domini 1416; et posteà fuit translatum ad hanc Capellam, in hoc tumulo reconditum cum corpore gloriosissimi Regis Domini Joannis, sui conjugis virtuosissimi, sub illâ formâ quæ in suo epitaphio continetur. Horum autem personas Deus omnipotens glorificare dignetur perpetuâ fœlicitate. Amen.

END OF DE SOUSA's HISTORY.

A Paper, of which the following is a Translation, I am indebted for to the Journal of the Right Honourable WILLIAM CONYNGHAM. *The Original is in French, and appears to have been written by one of the Fathers of Batalha for the above Gentleman when he visited the Monastery in 1783.*

AN ACCOUNT
OF THE
MODERN ESTABLISHMENT
OF THE
ROYAL CONVENT OF BATALHA.

THE number of Friars refiding in the Convent are forty-four, of the Dominican order; to wit, twenty-five in facred orders; two Deacons; four Novices; and thirteen Lay-brothers. They are governed by a Prior and three fubordinate dignitaries, viz. a Rector of Novices, a Vicar, and a Mafter of Morals. There are two Profeffors for teaching grammar to feculars, and another for inftructing them to read and write. The other officers of the Monaftery are, the *Sacrift, Precentor, Cellerarius, Gramatarius*, and the *Eleemofynarius*. There are alfo two Treafurers under the direction of the Prior; each has a feparate key of the cheft, which contains the ftock of the community.

The annual revenue of the Convent is computed at between ten and twelve thoufand cruzados [*], according to the fale of the fruit. The fixed revenue is 3,000 cruzados and forty *moyas* [†] of wheat, befides 200 *mil reis* received annually from the cuftom-houfe of Oporto.

The difburfements amount on an average to 7,000 cruzados a year. Each Friar is allowed 4,800 *reis* [‡] for his clothing. The tillage of the eftate called *Quinta da Vorgea* amounts to 400 *mil reis* per annum; at prefent it is farmed for one half of its produce. There are alfo let out four mills, one of flour and three of oil. In rebuilding the Dominican Church at Lifbon, which had been deftroyed by the great earthquake, the Convent contributes 300 *mil reis* annually. The expences attending the Church, in wax, &c. amount to 200 *mil reis*. The remainder of the income is expended in repairs and other contingencies.

There are fourteen fervants belonging to the Monaftery. The cook is allowed 4,800 *reis* a year, and wine at difcretion. The carters have but three moidores a year without victuals, except two, who are allowed four moidores. The fhepherd and fwineherd have each 600 *reis* a month, and four *alqueires* of maize. The two boys who attend the Sacrifty and Choir have no fixed falary.

Every year the Convent has four feafts and two days of double allowance; the ordinary allowance of each Father is 1½ lb. of meat, and the fame quantity of fifh; befides wine, fruit, &c.

[*] A cruzado (vello) is worth 2s. 3d.   [†] A moyo contains about 21½ bufhels.   [‡] A thoufand reis is equal to 5s. 7½d.

OF THE ROYAL MONASTERY OF BATALHA. 59

A Description of the several Parts of the General Plan.

	Feet.	In.
THE length of the Building from the Western entrance at A to the Eastern extremity at Z	416	7¼
The extent of the Church and Monastery from Y to B, North and South	541	0½

C The Mausoleum of King John the First.
D The Church.
E The Transept.
a Stairs leading to the Roof of the Church.
b Stairs leading to the Organ-loft, Roof of North Aisle, &c.
c The situation of the Organ.
d The Pulpit.
F An Altar dedicated to our Saviour.
G The Choir.
e The High Altar dedicated to GOD.
H Chapel of our Lady of the Assumption.
I Chapel of our Lady of the Rosary.
J Chapel of our Lady of Mercy.
K Chapel of St. Michael.
L The Sacristy.
M The Tower over which the Spire is built.
N *Caza da Prata*, or Room where the Plate, Relicks, &c. are deposited.
O Chapter-house.
P Loggia of the Mausoleum of King Emanuel.
Q Recesses for Altars.
f. f Small Recesses to contain the Phials with the Elements used in the Celebration of the Holy Sacrament.
R Mausoleum of King Emanuel.
S Chapels.
T The Chapel intended for King Emanuel.
g Apparently intended as a Repository for the Vestments of the Clergy.
V Excavations evidently intended for Sepulchres.
h The Entrance to the Excavations, which is filled up with hewn Stone without Cement.
i Stairs leading to the Platform over the Chapels.
W The Royal Cloister.
W W The Walks of the Cloister, through which the Friars and Novices pass in Procession from the Choir to the Refectory and back again, chanting grace before and after dinner *.
X A Garden.
k Cistern.
A Great Fountain.
B Refectory.
l Prior's Seat.
m The Pulpit, where one of the Novices reads the Holy Scripture whilst the Friars are at their meals.

n Apertures where two of the Novices receive the dishes from the Cook, which they carry to the Friars.
C The Kitchen.
D The Larder.
E A small Refectory, where the Prior and some of the principal Fathers occasionally dine.
F The Wine and Fruit Cellar.
G The Hall where the Professors give Lectures.
H The Sacrist's Store-room.
o The Belfry.
p Stairs leading to Prior's Apartments, &c.
I An open Arcade, where the Laundresses, &c. belonging to the Convent are permitted to enter.
J The School.
q Stairs leading to the Wheel of the Lay Brothers' Corridor.
K Servants' Hall.
r Stairs leading to the principal Cells.
s Corridor leading to the Church.
f Wax Chandler's Room.
L Servants' Cloister.
M A Court planted with Orange-trees.
N Lay Brothers' Cloister.
O A Garden.
P Novices' Cloister.
Q Novices' Garden.
R Store for Fuel.
S Oil Magazine.
T Wine Press.
t t Labourers' Store-rooms.
v v Artificers' Store-rooms.
V *Cloaca*.
w Stairs leading from the Dormitories to the Refectory.
x Servants' Cell.
W Magazines for provisions.
X Granary.
y Corridor.
z Stairs leading to Lay Brothers' Cells.
Y Labourers' Apartments.
Z Prior's Coach-house and Stable.
a a Stalls for Cattle.
b b Repository for Implements of Husbandry.
I. II. III. IV. Comprehends the space finished during the life-time of the Founder.
I. II. VI. V. Comprehends the space built by King *Edward*, Son of the Founder.

References to the Tombs.

a Tomb of King John the First and Queen Philippa his Consort.
b Tomb of *Don Pedro*, Duke of Coimbra, Knight of the Garter, &c.
c Tomb of *Don Anrique*, Duke of Viseu, Knight of the Garter, &c.
d Tomb of Don John, Master of the Order of *Santingo*, &c.

* The Novices walk first, and the Prior last, in the procession coming from the Refectory. In going thither the contrary order is observed.

e Tomb of the Infante *Don Fernando.*
  N. B. The above Princes were Sons of King John the First.
f Tomb of King *Edward* and Queen *Lianor.*
g Tomb of King John the Second.
h Tomb of *Don Lopo Diaz de Sousa* *.
i Tomb of *Dona Mecia,* wife of the *Conde de Miranda, Anrique de Sousa.*
k Tomb of Dona Isabel, Queen of Alfonso the Fifth.
l Tomb of a Cardinal whose name and family are unknown.
m Tomb of King Alfonso V.
n Tomb of Prince Alfonso, son of John II.
o A Tomb-stone on which is carved in relief a large Gothic D, surrounded with a glory. It is thought to be the Tomb of *Don Diogo Gonsalvez Travessos,* a great favourite of King John I.
p A plain Tomb-stone, without any inscription, under which, it is said, lies a Soldier who saved King John's life in battle.

Near the entrance of the Church are the Names of the following Workmen, who are there interred.

First Master Workmen.

Mestre Matheos. Portuguese. 1515.
Mestre Congiate. A stranger.
Mestre Conrado. A stranger.

First Master Workmen for the Windows.

Mestre Ugado. A stranger.
Mestre Whitaker. A stranger.

Ornaments, Mottos, &c. appertaining to the Royal Monastery of Batalha.

Figures 1, 2, 3, and 4. Capitals in the Nave of the Church.
Fig. 5. A Capital in one of the Windows. North Front.
6. ———— in the Arcade of the Royal Cloister.
7. ———— in one of the Windows. South Front.
8. ———— in the Mausoleum of the Founder.
9. ———— in the Western Porch.
10. ———— in the Transept entrance.
11. A Corbel supporting the Precentor's seat in the Choir.
12. Ornaments appended to the Intrados of the Arches in the Mausoleum of the Founder.
13. An Ornament in one of the Arches of the great Fountain.
14. A Specimen of the Ornaments placed on the Arches of contrary Flexion. West Front.
15. One of the Ornaments on the Angles of the Spire.
16. Profile of the above Ornament.
17 and 18. Pateras of the Pendent Orbs in the Chapter House.
19. A mural Font in the Mausoleum of the Founder.
20. A Patera in the centre Vault of the above Mausoleum.
21. A Patera in one of the Aisle Vaults of the Church.
22. A Figure supporting one of the Ribs of the Vault in the Chapter House, supposed to represent the Architect of that Fabrick.
23. John the First's Sword and Battle-axe.
24. John the First's Helmet.
25. John the Second's Sword and Battle-axe.
26. Ornaments of the Mausoleum of the Founder, externally.
27. A Tablet in the front of a small House opposite to the Church, said to be wrought by the Workmen of Batalha.
No. 28. Motto on the Tomb of King John the First.
No. 29. ———— on the Tomb of Don Pedro.
No. 30. ———— on the Tomb of Don *Anrique.*
No. 31. ———— on the Tomb of Don John.

Columns, Ornaments, and Hieroglyphicks in the Mausoleum of King Emanuel.

Figures 1, 2, 3, 4, 5, and 6. Columns at the entrance of the Mausoleum.
Fig. 7. Pedestal of the above Columns.
8. In each Angle of the Octagon, there is a Column like this, from the Top of which the Ribs of the intended Vault commence.
9. Hieroglyphicks at the Entrance of the Mausoleum.
10. Ornaments in the intervening spaces of the Columns at the Entrance.
11. Architrave, Frieze, and Cornice of the Mausoleum. Internally.
12, 13, and 14. Vases on the Pedestals and Pilasters of the intended Orchestre.
15. An Ornament over the Loggia.
16. A Baluster in front of the intended Orchestre.
17. The Initials of the Founder's name, i. e. *Manoel Rex.* These letters are often repeated on the Architraves of the large Windows of the Octagon.

† This tomb is modern, but well designed and executed in different kinds of marble; and supported by three crouchant lions of Sienna, with their paws resting upon as many balls, intended, I suppose, to indicate mutability. The top of the monument is finished with a ducal coronet resting upon a cushion, and supported by two weeping figures of Carara marble. In the middle is a tablet, handsomely ornamented, bearing the following letters,
X · R · P · M · H · S · E
The meaning of these letters, according to the Almoner of the Convent, is as follows,
Decima, Regis, Persona, Masculina, Hic Sepulta Est.

OBSERVATION.

THE Work from which we have translated the foregoing Account of Batalha, entitled *Historia de St. Domingos*, was published at Lisbon in the year 1622, under the names of Cacegas and De Sousa. The former, who was historiographer to the Dominican Order in the Portuguese dominions, first compiled the History of that Order; and some years after it was continued by De Sousa, who also made many alterations in, and additions to, what his predecessor had written. From the superiority this writer possessed as an historian over Cacegas, and the approbation his labours received from the literati of Portugal, the Work at present bears the title of De Sousa's History of the Dominican Order.

From this History I have selected such facts as appeared essential to my purpose, and no more; and even the passages thus selected are not always rendered word for word with the original text. Where the author had either mistook or misunderstood the terms of architecture, which indeed is excusable in an historian, I thought myself authorised in giving them their proper names. I have also corrected the dimensions he assigns to some parts of the edifice when I found them wrong, and inserted the real measurements in their place, in feet and inches. As it would be tedious and uninteresting to enter upon a discussion with my Author on these points, I generally omitted to notice them. Such are the liberties taken with Father De Sousa's Account of Batalha in the foregoing Translation, of which I thought it necessary to apprise the Reader.

MEMOIRS OF FATHER LEWIS DE SOUSA.

AS the circumstance which induced our Author to seclude himself from the world and become a Friar is rather singular, a short account of it may not be unacceptable.

In 1578, when Don Sebastian King of Portugal was defeated and slain in a pitched battle against Muly Moloch Emperor of Morocco, many of the nobility of Portugal, who accompanied him, shared the same fate; and others who fell into the enemy's hands were made captives.

Amongst the gentlemen who accompanied King Sebastian in this unfortunate expedition, was one whose name the Biographer has omitted; it was included, however, in the return of the slain. When his wife, who resided in Lisbon, received the intelligence, she nevertheless entertained hopes that it might have been a mistake, and that Heaven would yet favour her with a sight of him.

Under this pleasing expectation she remained ten years, notwithstanding the repeated accounts she received from the agents employed to redeem the captives confirmed the relation of his death. Her friends, who were convinced of the truth of it, entreated her to relinquish the idea of ever seeing him, and to enter once more into the marriage state.

Sousa at this time moved in the first circles of fashion; his company was much sought for, as he was an excellent scholar, as well as an accomplished gentleman. He paid his addresses to this lady; her incredulity respecting her husband's death at this time began to give way, and she was prevailed on by her relations to give him her hand.

Accordingly they were married, and lived together in the greatest harmony. But it was of short duration: a merchant from Africa arrived in Lisbon, sought out the lady, and informed her, that he was charged with a commission from her husband, who was in captivity, and relied upon her affection to expedite his release.

The unfortunate woman, quite overwhelmed with shame and surprise, in this affecting dilemma asked De Sousa's advice, who was also astonished at the news. As he was a prudent and conscientious man, he resolved to be guided in a matter of such delicacy by the purest dictates of honour.

In the first place, in order to ascertain the fact, he had recourse to an ingenious expedient: he conducted the messenger to a picture-gallery in his house, and told him that a portrait of the gentleman whom he affirmed to have seen was in the collection, and requested him to point it out, as a proof that there was no mistake in his declaration. The merchant endeavoured to excuse himself, saying that a long state of servitude and cruel treatment had made such a change in the captive gentleman, that he doubted if his most intimate friends could recognize him if he were present: nevertheless, says he, some leading features induce me to think that this is his portrait, pointing to the identical one. Sousa from this and other collateral circumstances was now convinced of the truth of the whole, and applauded the merchant for his humanity.

This affair affected Sousa very much: he deliberated with himself in what manner to act; at length he resolved, having no children to provide for, to retire from the world and seclude himself in a monastery. The wife approved the resolution; and, as a proof of her grief and affection, retired also into a nunnery near Lisbon. But previous to their seclusion, they used every means in their power to rescue the unfortunate gentleman from captivity.

Sousa now entered into the Dominican Order, and lived in the Convent of Bemfica near Lisbon. The Fathers of this Order, desirous of completing the History of their Foundation, thought this a favourable opportunity; and knowing Sousa to be a man of great talents, they requested him to undertake the task, and perfect what Cacegas, a Friar of the same Order, had begun. He accordingly set about it, and, after many years labour, published it in 1619, under the name of Cacegas and his own; thus, from his extreme modesty, dividing the honour of the work, the whole of which he could justly claim as his own. But posterity has done justice to his memory, and Cacegas's name is now remembered only through De Sousa's works.

His facts are said to be accurate and well arranged; his deductions natural and solid; his style throughout is simple and nervous; and, what adds greater honour to his memory, he was a man of exemplary piety and humanity.

FINIS.

DIRECTIONS FOR PLACING THE PLATES.

THE Dedication to precede the Preface.
Plates 1, 2, 3, 4, Introduction; and the Tranfverfe Section of the Church; after the Introduction.

The reft of the Plates to be placed after the Hiftory, in the following Order:

General Plan.
North Elevation of the Church, &c.
Longitudinal Section of the Church.
Elevation of the Chancel.
Section of the Chapter Houfe.
Elevation of the Refectory.
Weft Elevation of the Church.
Elevation of the Tranfept Entrance.

Interior View of the Church.
Elevation of the Maufoleum of King John.
Section of the Maufoleum of King John.
Effigies of King John and Queen Philippa.
Entrance to the Maufoleum of King Emanuel.
Defign for completing the Maufoleum of K. Emanuel.
Arches appertaining to the Maufoleum of K. Emanuel.
Ornaments, Mottos, &c.
Columns, Ornaments, and Hieroglyphicks.
Elevation of the Pillars of the Church.
Elevation and Section of the Spire.
Rails, Cornices, and Arched Modillions.

   Total, Twenty-feven with the Title.

LIST OF SUBSCRIBERS.

HIS BRITANNIC MAJESTY.
HER MOST FAITHFUL MAJESTY.
HIS ROYAL HIGHNESS THE PRINCE OF BRAZIL, 10 sets.
HER ROYAL HIGHNESS THE PRINCESS OF BRAZIL.
HER ROYAL HIGHNESS THE PRINCESS DOWAGER OF BRAZIL.
HER ROYAL HIGHNESS THE PRINCESS SOPHIA OF GLOUCESTER.

A

CHEVALIER de Almeida, Portuguese Ambassador at London.
Viscount de Anadia, Portuguese Ambassador at Berlin.
Viscount de Asseca, Lisbon.
Mr. João Antaõ de Almada, Lisbon.
Monsenhor Arcinolli, Lisbon, 3 sets.
Don Bernardin Freire de Almada, Lisbon.
Earl of Aylesford.
Mr. Andrade.
Mr. Archer, Dublin.
Richard Amies, Esq. Lisbon.
John Armstrong, Esq. Architect.

B

Cardinal Bellisomi, Nuncio at Lisbon.
Bishop of Beja, Portugal.
Bishop of Castello Branco, Portugal.
Sir Joseph Banks, Bart. P. R. S. F. S. A. &c.
Owen Salusbury Brereton, Esq. F. R. S. F. S. A.
William Beckford, Esq.
William Bosville, Esq. F. R. S.
John Thomas Batt, Esq.
Richard Bullock, Esq.
Richard Buller, Esq.
George Byfield, Esq. Architect.
Mr. John Berril, Dublin.
Mr. Richard Butler, Dublin.
Richard Burn, Esq. Lisbon, 2 sets.
Francis Beasley, Esq. Oporto.
William Boys, Esq.
Charles Beazley, Esq. Architect.
Sir Francis Blake, Bart.
Edward Burton, Esq.
Thomas Barrett, Esq.
Charles Bulfinch, Esq. Boston, America.
Col. Biefster, Lisbon.

C

Right Honourable William Conyngham, M. R. I. A. F. S. A. &c. 10 sets.
Lord Viscount Conyngham.
Lord Carnelford.
Bishop of Cloyne.
R. Pole Carew, Esq. M. P.
John Clements, Esq.
Sir William Chambers, K. P. S. F. R. S. R. A. &c. Architect to the Queen.
Charles Cameron, Esq. Petersburg, Architect to the Empress of Russia, 2 sets.
John Haydon Cardew, Esq.
Thomas Collins, Esq. M. P. R. M. T.
Messrs. Colnaghi and Co.
John Coltsman, Esq. Lisbon.
Denis Connell, jun. Esq. Oporto.
R. M. T. Chifwell, Esq. M. P.
George Cooke, Esq.
John Claxton, Esq. F. A. S.
Mr. W. Cole, Architect.
Mr. Crace.
William Carr, Esq. Architect.
Senhor João Vidal da Costa, Portugal.

D

Lord Viscount Dillon.
George Dance, Esq. Architect to the City of London.
Gerard Devisme, Esq.
Mr. Timothy Doyle, Dublin.
Mr. Whitmore Davis, Architect, Dublin.
George Dohrman, Esq. Lisbon.
William Danby, Esq.
Samuel Dobrée, Esq.

E

Earl of Exeter.
Count de Ega, Lisbon.
Mr. John Edwards.
Rev. John Eyre.
Mr. Elmsley.

F

Don Miguel Frojaz, Lisbon.
Monsenhor Freire, Lisbon.
William Fauquier, Esq.
Mr. Fry.
Mr. Thomas Frazer.
Mr. John Foster, jun. Architect.
Dr. Antonio Mendes Franco, Lisbon.

G

Count Gervazone, Lisbon.
Richard Gough, Esq. F. R. S. F. S. A.
Dr. Gray, R. S.
Sir Richard St. George, Bart.
Mr. Thomas Gowland.
William Gonne, Esq. Lisbon.
Walter Grosset, Esq. Lisbon.
Timothy Goodall, Esq. Lisbon.
Abbé Garnier, Chaplain to the French Factory, Lisbon, 2 sets.
George Gibson, Esq. Architect.
John Groves, jun. Esq. Architect.

H

Lord Howard.
Sir John Hort, Bart.
Sir James Hall, Bart.
Sir Richard Hoare, Bart.
Rev. Herbert Hill, M. A. Chaplain to the British Factory, Lisbon, 2 sets.
Mrs. Harcourt.
John Heathcote, Esq.
Adair Hawkins, Esq.
Capt. Hanchet.
Charles Hoare, Esq.
Mr. John Hartwell, Dublin.
Mr. Daniel Harris, Architect.
Richard Holland, Esq. Architect.
Mr. John Josiah Holford.
George Harding, Esq. M. P.
Benjamin Harrison, Esq. Lisbon.
Peter Henry Henrichsen, Esq. Lisbon.
Mr. John Hurst.

J

John James, Esq.
Colonel Johnes, M. P.
Rev. T. Jones, M. A.
Joseph Johnston, Esq.
Bromley Illius, Esq. Lisbon.
Richard Jupp, Esq. Architect.

K

Edward King, Esq. F. R. S. and F. S. A.
J. T. Koster, Esq. Lisbon, 2 sets.
L. A. Kantzon, Esq. his Swedish Majesty's Agent at the Court of Lisbon.

L

Marquis of Lansdown.
Earl of Leicester, P. S. A. and F. R. S.
Lord Lismore.
William Lock, Esq. F. R. S. and F. S. A.
Charles Lambert, Esq. F. S. A.
Thomas Lane, Esq.

LIST OF SUBSCRIBERS.

Don Lourenço de Lencaftre, Lifbon.
Senhor João Gabriel Lobo, Lifbon.

M
Marquis das Minhas, Lifbon, 4 fets.
Marquis de Marialva, Lifbon.
Marquis de Caftello Melhor, Lifbon.
Earl of Mount Edgcumbe.
Earl of Moira.
Countefs of Moira.
Charles Marfh, Efq. F. S. A.
William Meredith, Efq.
—— Mc Gowan, Efq.
Jofeph May, Efq.
Robert Mitchell, Efq. Architect.
Mr. Michael Murphy, Dublin.
Mr. Richard Morrifon, Architect, Dublin.
Charles Murray, Efq. Conful General at Madeira.
Senhor Jozé Caetano Machado, Lifbon.
John Henry Metzener, Efq. Lifbon.
Mr. T. Milton.
Jofeph Mufgrave, Efq.
Henry Muirs, Efq.
Senhor Jozé de Cotta Mourão, Lifbon.

N
Duke of Norfolk, 4 fets.
Duke of Northumberland, 2 fets.
Hon. Mrs. O'Neale.
Sir Roger Newdigate, Bart.
Mr. George Nicols.
William Nafh, Efq. Oporto.
John Nafh, Efq. Architect.
William Northey, Efq.
Rev. P. Newcombe.

O
Earl of Orford.
Earl of Upper Offory.
John Frederic Oftervald, Efq. *Chargé d'Affaires, Lifbon.*
Rev. Hugh Owen.

P
Marquis de Pombal, Lifbon, 6 fets.
Chevalier Pinto, Secretary of State, Lifbon, 6 fets.
Don Bafilio Francifco de Carvalho Pinto, Malta.
John Peachy, Efq. F. R. S.
Roger Palmer, Efq.
Thomas Penuant, Efq.
Mr. John Plaw, Architect.
Mr. Thomas Payne.
William Porden, Efq. Architect.
Mr. Pilkington.
Thomas Powell, Efq.

Q
Don Pedro Joaquim Quintella, Lifbon.
Doctor Quin, Dublin.

R
Sir George Robinfon.
William Reveley, Efq. Architect.
Alvez Rebello, Efq.

S
Earl Spencer.
Chev. de Soufa, Portuguefe Ambaffador at Stockholm.
Chev. de Soufa, Portuguefe Ambaffador at Copenhagen.
William Seward, Efq. F. R. S. and F. S. A.
Hugh Skeys, Efq. Dublin, 2 fets.
Sir Richard St. George, Dublin.

Thomas Sikes, Efq.
Senhor Jozé Penheiro Salgado.
Captain Smith.
John Soane, Efq. Architect to the Bank of England.
Mr. Dias Santos.
Mr. Roiz de Saá.
Senhor Jozé Correa de Saá, Lifbon.
Senhor João Correa de Saá, Lifbon.
John Searle, Efq. Oporto.
Nicholas Bernard Schick, Efq. Lifbon.
Jofeph Sill, Efq. Lifbon.
Richard Sealy, Efq. Lifbon.
Senhor Manoel Caetano de Soufa, Lifbon.
William Stephens, Efq. Marinha Grande, Portugal, 16 fets.
Senhor João Vidal de Cofta e Soufa, Lifbon.
Mr. Alexander Stevens, Architect.
Sir Chriftopher Sykes, Bart.
Senhor Jozé Antonio da Sylveira, Lifbon.
Lourenço Rodrigues de Sá, Efq.

T
John Topham, Efq. F. S. A.
—— Todd, Efq.
Richard Townfend, Efq. M. D. Dublin.
Mr. Richard Tawney, Architect.
Mr. W. Todd.

V
Count de St. Vincent, Lifbon.
Mr. John Upward.
Timothy Verdier, Efq. Lifbon.

W
Hon. Robert Walpole, Britifh Envoy, Lifbon.
Jofeph Windham, Efq. F. S. A.
William Watfon, Efq. F. R. S.
Mifs Walker.
William Webb, Efq.
Dr. Willis.
John Woodhoufe, Efq.
Daniel Whalley, Efq.
Benjamin Weft, Efq. P. R. A. Painter to his Majefty.
James Wyatt, Efq. R. A. Architect to his Majefty.
Mr. Thomas Whetten, Architect.
Jofeph Cooper Walker, Efq. M. R. I. A. Dublin.
William Whitehead, Efq. Britifh Conful at Oporto.
William Warre, Efq. Oporto.
Mr. Thomas Wright.
Mr. Watfon.
Mr. Jeffery Wyatt.
Mr. Waters.

Y
Hon. John Yorke.
Sir George Young, Bart. F. R. S.

Colleges and Libraries.
His Majefty's Library.
Antiquary Society.
Royal Academy.
Britifh Mufeum.
Trinity College, Dublin.
Royal Irifh Academy.
Dublin Society.
Royal Convent of Batalha.
Univerfity of Coimbra.
Leeds Library.
Charles Town Library, South Carolina.
Royal Library, Gottingen.

ADDITIONAL NOTE.

Since the former fheets of this work were printed, the author has been favoured by the Rev. Herbert Hill, chaplain to the Britifh Factory at Lifbon, with an Extract from a Portuguefe Hiftorian *; wherein are afcertained, apparently from good authority, the name and country of the Architect of BATALHA. The following is a Tranflation of the paffage:

" Fr. Luis de Soufa, in the Hiftory of the Dominican Order, part I. and D. Fernandes de Menezes, Count de Ericeira, at " the end of the Life of King John I. have both defcribed the Royal Monaftery of Batalha with all the exactnefs and elegance " which it well merits. To thefe Authors I refer the Reader for an account of that noble edifice; and left any unpolifhed lan-
" guage fhould fully its renown, I fhall only obferve, that the Architect of it was an *Irifhman*, named *David Hacket*, who then
" lived in Vianna da Caminha, as may be feen in one of the Memoirs of Fr. Antonio da Madureira, a Dominican friar, and a
" celebrated genealogift."

* " *Jens Bures de Sylva Adm. del Rey D. Jozé 1º. Tom. 2. p. 133.*"

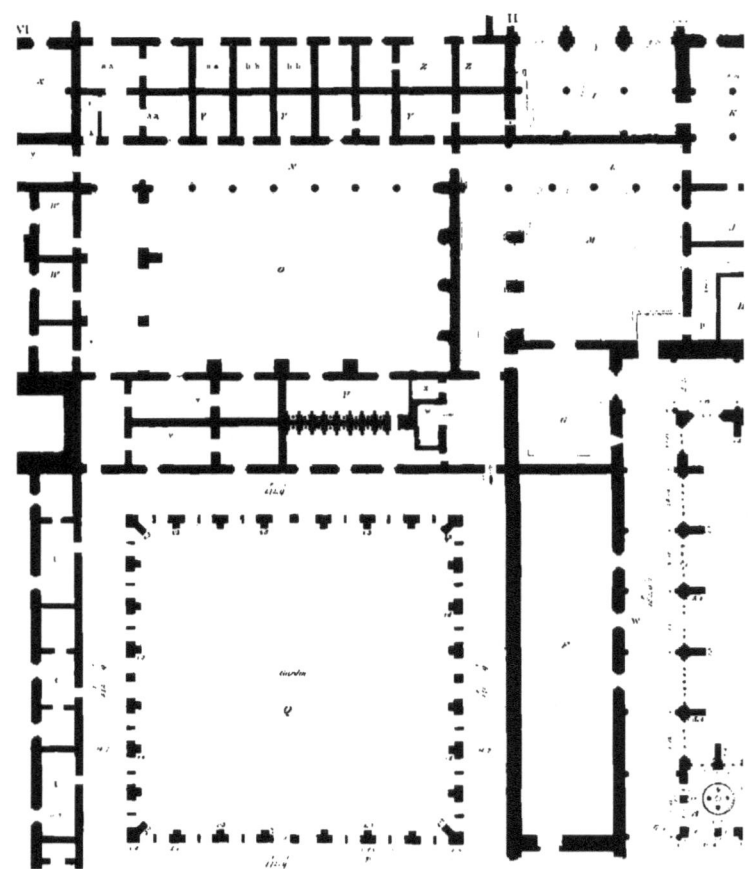

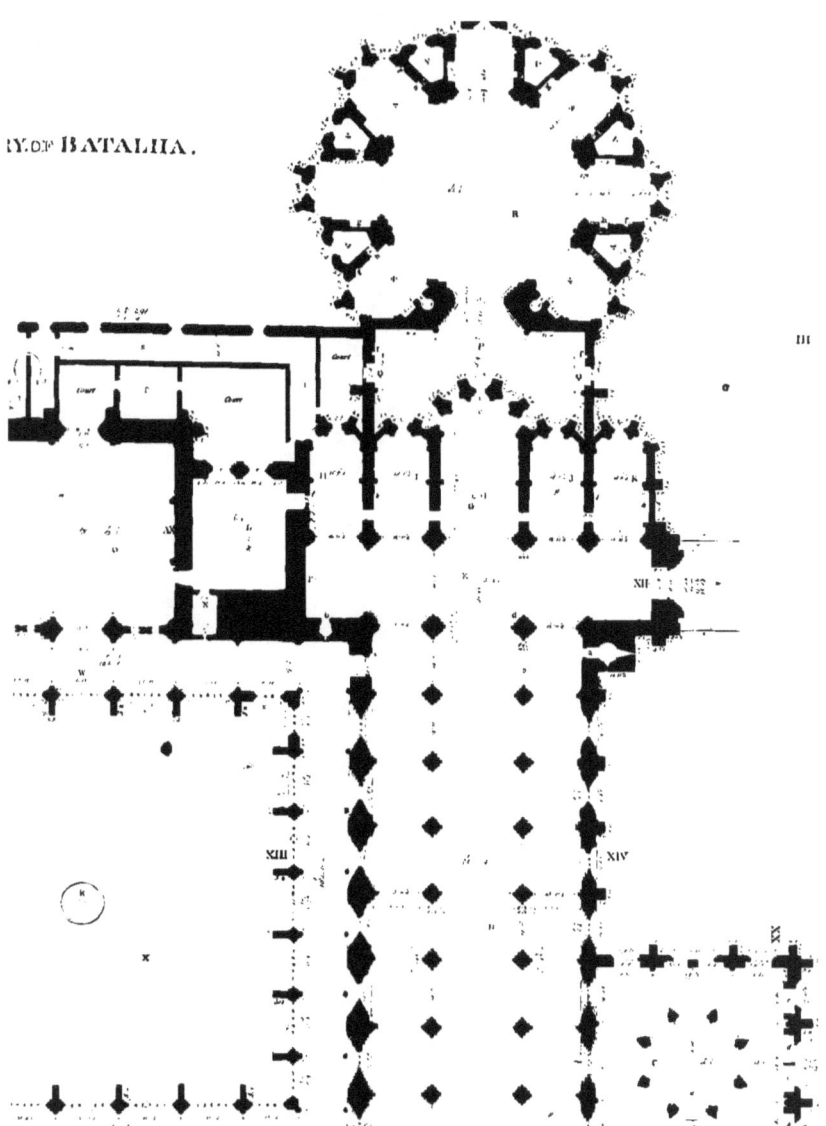

THE NORTH ELEVATION OF THE CHURCH OF B.

A, WITH THE UNFINISHED MAUSOLEUM OF KING EMANUEL.

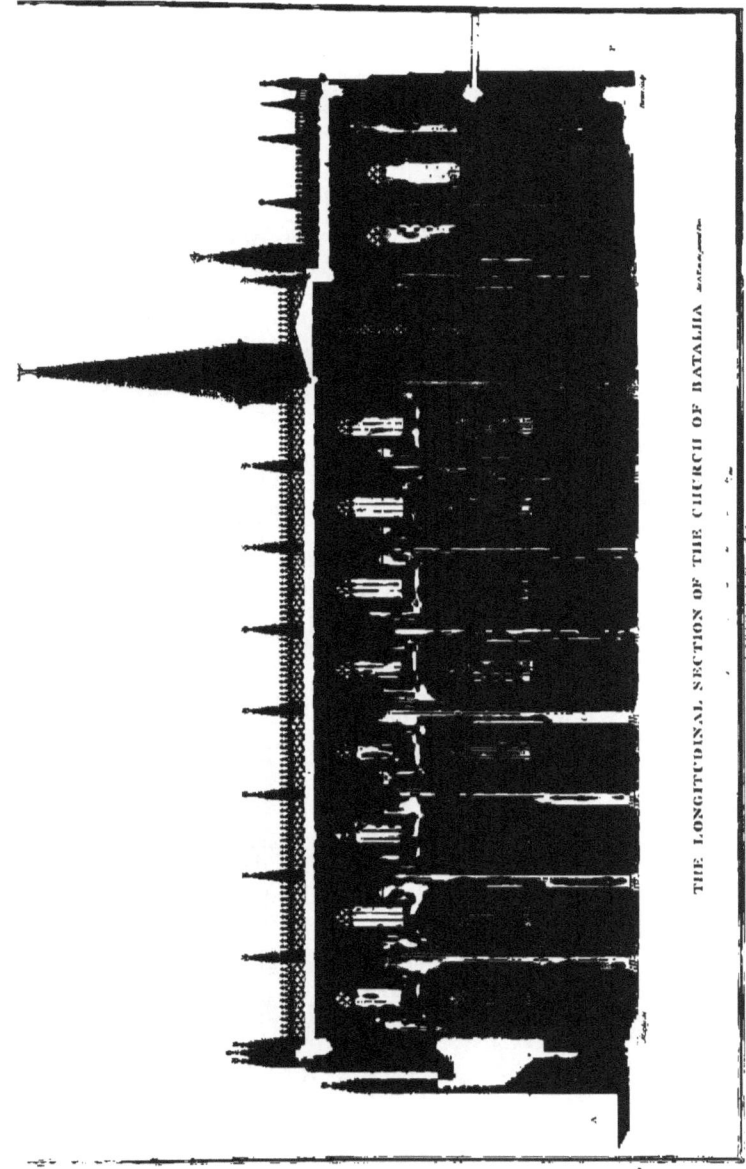

THE LONGITUDINAL SECTION OF THE CHURCH OF BATALHA

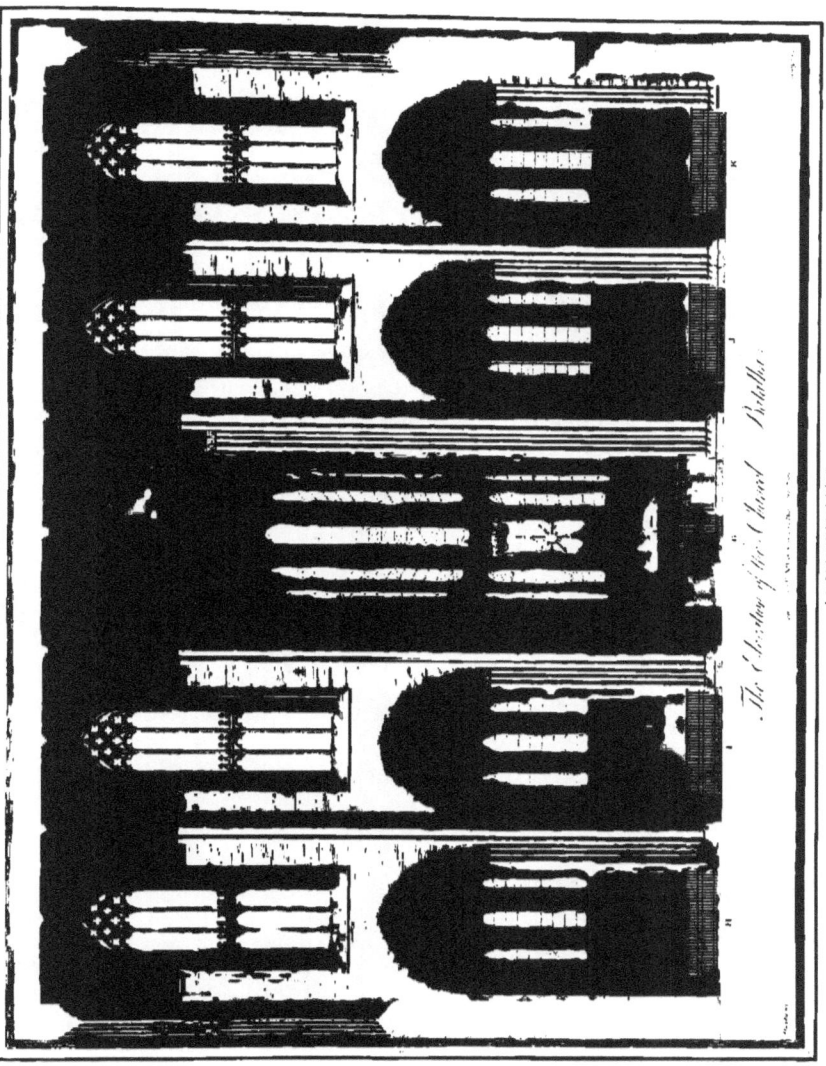

A SECTION OF THE CHAPTER HOUSE AT BATALHA

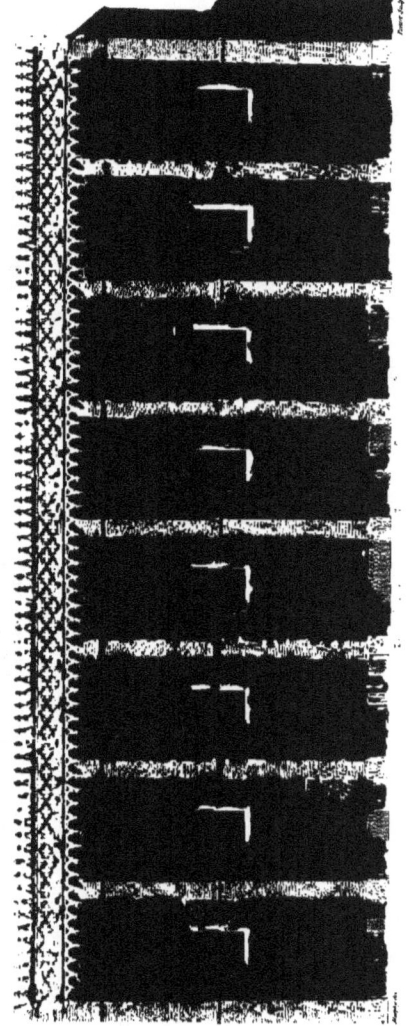

THE WEST ELEVATION OF THE REFECTORY OF BATALHA.

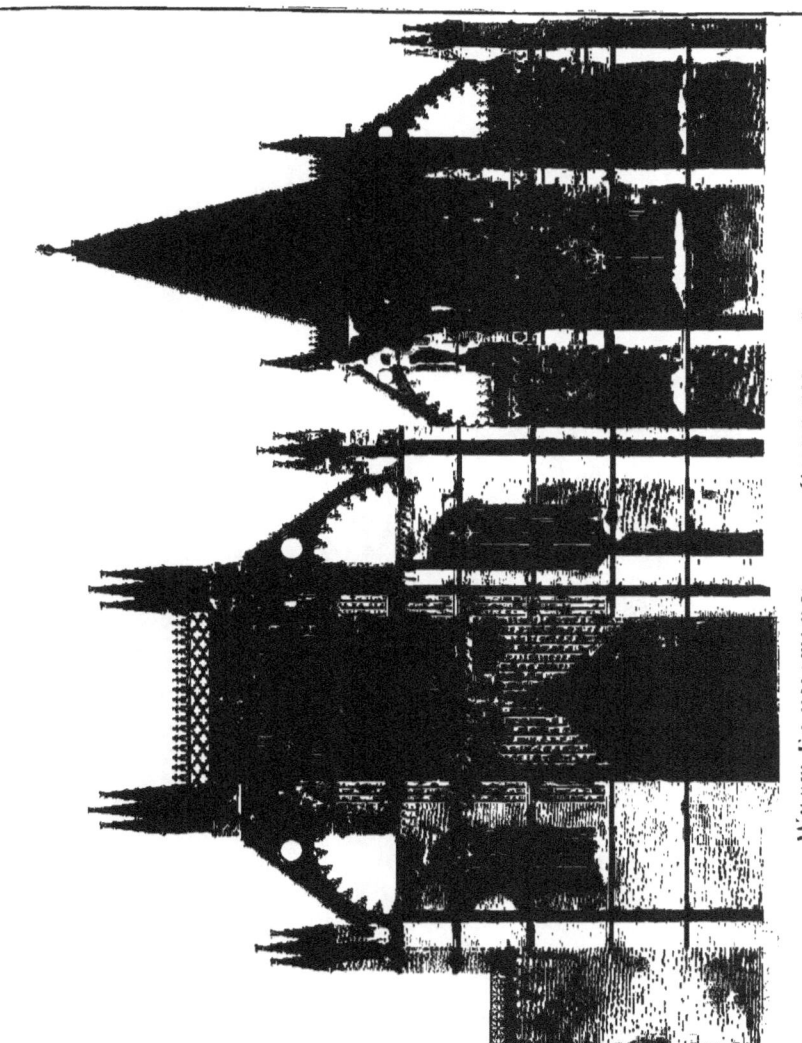

WEST ELEVATION OF THE CHURCH. BATALHA.

ELEVATION OF THE TRANSEPT ENTRANCE. BATALHA.

INTERIOR VIEW OF THE CHURCH OF BATALHA.

THE SOUTH ELEVATION OF THE MAUSOLEUM OF KING JOHN 1ST AT BATALHA.

SECTION OF THE MAUSOLEUM OF KING JOHN 1.

THE EFFIGIES OF KING JOHN THE FIRST AND QUEEN PHILIPPA.

DESIGN FOR COMPLETING THE MAUSOLEUM OF KING EMANUEL

ARCHES APPERTAINING TO THE MAUSOLEUM OF K. EMANUEL, BATALHA

ORNAMENTS MOTTOS &c APPERTAINING TO THE ROYAL MONASTERY OF BATALHA.

COLUMNS, ORNAMENTS, AND HIEROGLYPHICKS, IN THE MAUSOLEUM OF KING EMANUEL.

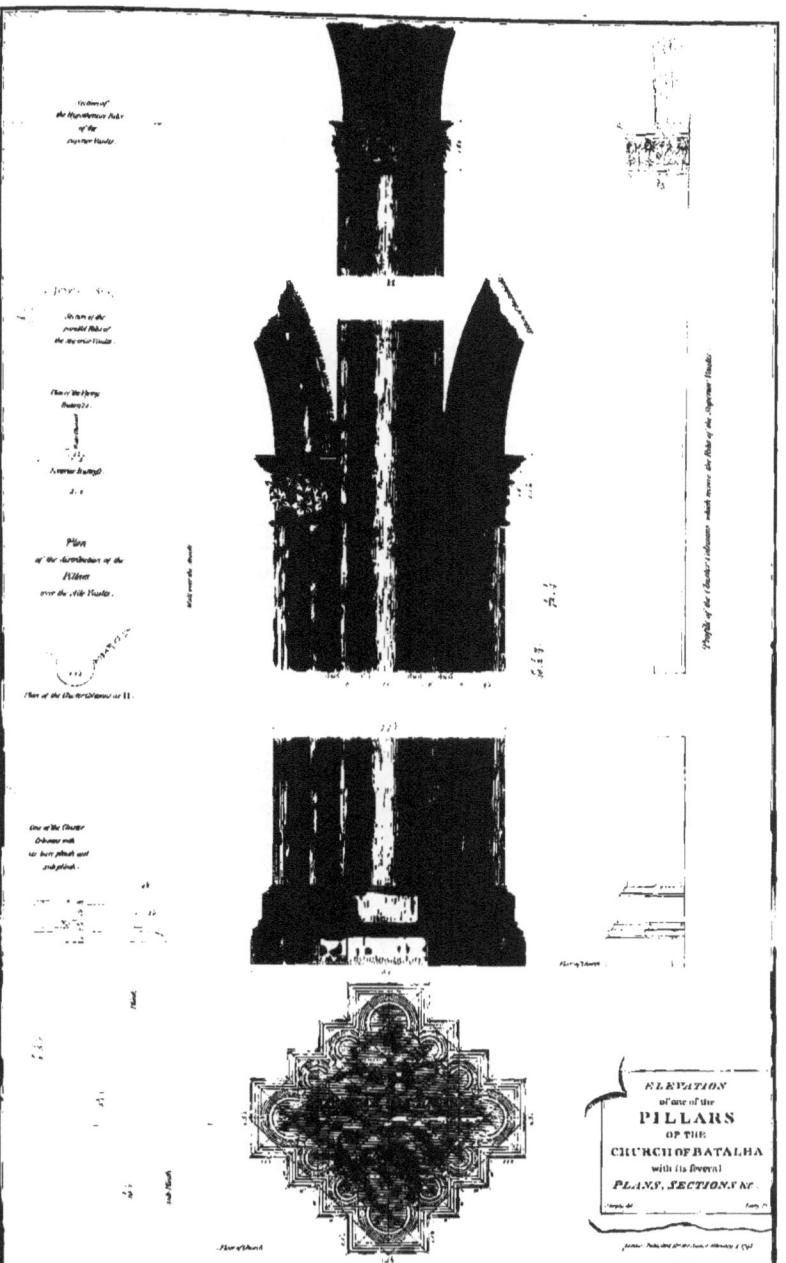

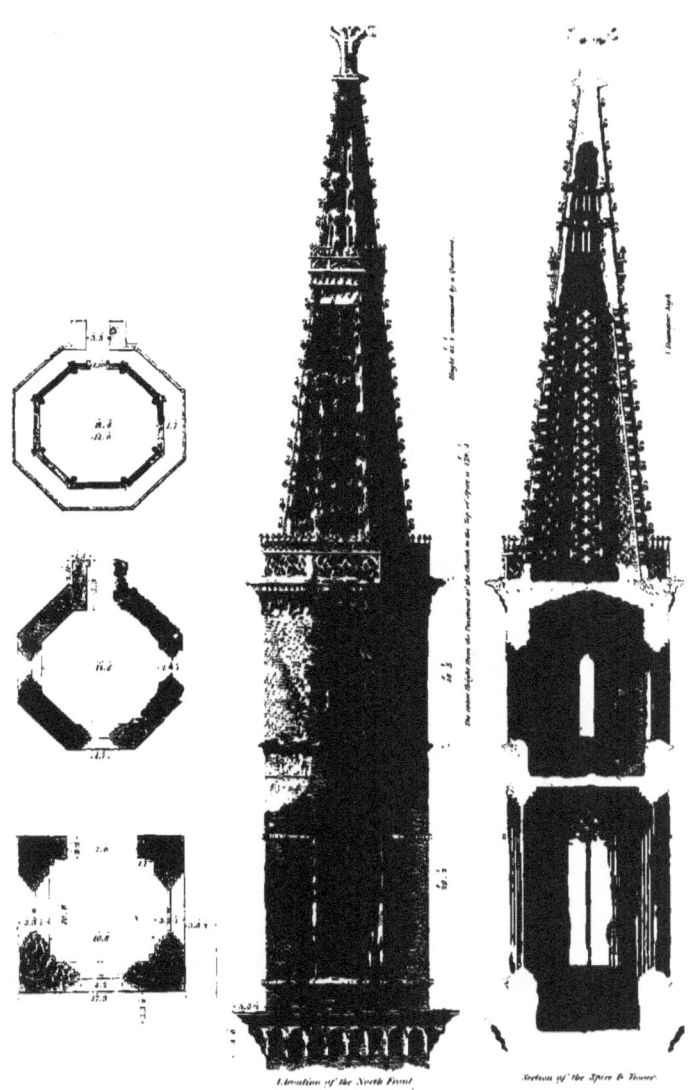

THE SPIRE OF THE NORTH END OF TRANSEPT_BATALHA.

RAILS · CORNICES · AND · ARCHED · MODILLONS: BATALHA.

A CATALOGUE of Modern Books on *Architecture*,
THEORETICAL, PRACTICAL, AND ORNAMENTAL;
VIZ.
Books of Plans and Elevations for Houses, Temples, Bridges, &c.
Of Ornaments for Internal Decorations, Foliage for Carvers, &c.
On Perspective.
Books of Use to Carpenters, Bricklayers, and Workmen in general, &c. &c.

Which, with the best Ancient Authors, are constantly on Sale

At I. and J. Taylor's Architectural Library, No. 56, High Holborn, London:
Where may be had the Works of the most celebrated *French Architects* and *Engineers*.

PLANS, Elevations, Sections and Views of the Church of *Botalha*, in the Province of *Estremadura*, in Portugal, with an History and Description by Fr. Luis de S ouſa, with Remarks; to which is prefixed, An Introductory Diſcourſe on the Principles of *Gothic Architecture*, by *James Murphy*, Architect. Illustrated with 27 plates. Elegantly printed on imperial folio, and hot pressed, price 4l. 14s. 6d.

The *Architecture* of this Structure is of the best Gothic, and one of the most chaste and magnificent specimens of the style existing; and well deserves the attention of the Antiquary and the Artist, for its purity of design and elegance of ornament.

THE Ancient Buildings of Rome, accurately measured and delineated, by *Antony Desgodetz*, with Explanations in French and English; the Text translated, and the Plates engraved, by the late *Mr. George Marshall*, Architect, 2 vols. imperial folio, with 137 plates, price 5l. 5s. half-bound, or 4l. 14s. 6d. sewed.

Desgodetz's Antiquities has ever been highly valued by Amateurs and Professors, for the Accuracy of the Admeasurements, and the Choice of elegant Subjects. The Work includes Designs at large of the following celebrated Structures, viz. the Pantheon, Temple of Bacchus, Temple of Vesta at Rome; Temple of Vesta at Tivoli; Temple of Fortuna Virilis; Temple of Antoninus and Faustina; Temple of Concord; Temple of Jupiter Stator; Temple of Jupiter Tonans; Temple of Mars the Avenger; Peristyle of Nero; Basilica of Antoninus; Forum of Nerva; Portico of Septimius Severus; Arch of Titus; Arch of Septimius; Arch of the Goldsmiths; Arch of Constantine; The Colisaeum; The Amphitheatre at Verona; The Theatre of Marcellus; The Baths of Diocletian; and The Baths of Paulus Emilius.

Thirty Capitals of Columns, with Six Friezes, from the Antique. Engraved in Aquatinta by *G. Richardson*, in 30 Plates. Quarto, 15s.

Designs for *Shop-Fronts* and *Door-Cafes*, on 27 plates, quarto, 10s. 6d.

Designs for *Monuments*, including *Grave-Stones*, *Comportments*, *Wall-Pieces*, and *Tombs*. Elegantly engraved on 40 quarto Plates, half-bound, 16s.

Designs for *Chimney Pieces*, with Mouldings and Bases at large, on 19 quarto Plates, 10s. 6d.

The *Rudiments of Ancient Architecture*: containing an Historical Account of the Five Orders, with their Proportion, and Examples of each from Antiquity: Also, Extracts from *Vitruvius*, *Pliny*, &c. relative to the Buildings of the Ancients; calculated for the Use of those who wish to attain a Summary Knowledge of the Science of Architecture; with a Dictionary of Terms. Illustrated with eleven Plates, boards, 6s.

Sketches for Cottages, Villas, &c. with their Plans and appropriate Scenery, by *John Soane*; to which is added, Six Designs for improving and embellishing Grounds, with Explanations, by an *Amateur*, on 54 Plates elegantly engraved in aquatinta, 2l. 12s. 6d. half bound.

Plans, Elevations, and Sections, of Buildings, executed in the counties of Norfolk, Suffolk, Yorkshire, Wiltshire, Warwickshire, Staffordshire, Somersetshire, &c. By *John Soane*, Architect. On forty-seven folio plates, 2l. 2s.

The Come on imperial paper, 5l. 15s. 6d.

Plans, Elevations, and Sections, of Noblemen and Gentlemen's Houses, Stabling, Bridges public and private, Temples, and other Garden Buildings, executed in the Counties of Derby, Durham, Middlesex, Northumberland, Nottinghamshire, York, Essex, Wilts, Hertford, Suffolk, Salop and surry; by *James Paine* Architect, Joint Architect in the Board of Works: 2 vols. with 176 very large folio Plates, 6l. 16s. 6d. half-bound.

The Designs of *Inigo Jones*, consisting of Plans and Elevations for Public and Private Buildings; including the detail of the intended Palace at Whitehall; published by *W. Kent*, with some additional Designs, 2 vols. imperial folio, 4l. 4s. in sheets; or half-bound, 4l. 14s.

Plans, Elevations, and Sections, of the *House of Correction* for the County of *Middlesex*, erected in Cold Bath Fields, London; together with the Particulars of the several Materials to be contracted for, and manner of doing the same in building. N. B. This Work is engraved from the original designs, and published with the authority of the Magistrates; by *Charles Middleton*, Architect. Engraved on 33 plates, imperial folio, half bound, 1l. 11s. 6d.

The Cabinet-Maker and Upholsterer's *Guide*; or Repository of Designs for every article of Household Furniture, in the newest and most approved taste; displaying a great variety of patterns for Chairs, Stools, Sofas, Confidantes, Duchesse, Side Boards, Pedestals and Vases, Cellerets, Knife Cases, Desk and Book Cases, Secretary and Book Cases, Library Cases, Library Tables, Reading Desks, Chests of Drawers, Urn Stands, Tea Caddies, Tea Trays, Card Tables, Pier Tables, Pembroke Tables, Turn-over Tables, Dressing Glasses, Dressing Tables and Drawers, Commodes, Rudd's Tables, Bidets, Night Tables, Basin Stands, Wardrobes, Pot Cupboards, Brackets, Hanging Shelves, Fire Screens, Beds, Field Beds, Sweep Tops for ditto, Bed Pillars, Candle Stands, Lamps, Pier Glasses, Frames for Busts, Cornices for Library Cases, Wardrobes, &c. at large. Ornamented Tops for Pier Tables, Pembroke Tables, Commodes, &c. &c. in the plainest and most enriched styles, with a scale to each, and an explanation in letter-press. Also the Plan of a Room, shewing the proper distribution of the furniture. The whole exhibiting near three hundred different designs, engraved on one hundred and twenty-eight folio plates: from drawings by *A. Hepplewhite* & Co. Cabinet-Makers, bound, 2l. 2s.

The Builder's Price Book; containing a correct *List of the Prices allowed by the most eminent Surveyors in London to the several artificers concerned in building*; including the Journeymen's Prices. A new edition, corrected, with great additions, by an experienced surveyor, sewed, 2s. 6d.

Ferme Ornée, or Rural Improvement; a Series of Domestic and Ornamental Designs, suited to Parks, Plantations, Rides, Walks, Rivers, Farms, &c. consisting of Fences, Paddock Houses, a Bath, Dog-Kennels, Pavilions, Farm-Yards, Fishing-Houses, Sporting-Boxes, Shooting Lodges, Single and Double Cottages, &c. calculated for Landscape and Picturesque Effects. By *John Plaw*, Architect, engraved in aquatinta on 38 plates, with appropriate Scenery, Plans, and Explanations, quarto, boards, 1l. 11s. 6d.

Rural Architecture, or Designs from the simple Cottage to the Decorated Villa, including some which have been executed, by *John Plaw*, on 62 Plates, with Scenery, in aquatinta, in boards, 2l. 2s.

Familiar Architecture; consisting of original Designs of Houses for Gentlemen and Tradesmen, Parsonages and Summer Retreats; with Back-Fronts, Sections, &c. together with Banqueting-Rooms, and Churches. To which is added, the Masonry of the Semicircular and Elliptical Arches, with practical Remarks. By the late *Thomas Rawlins*, Architect. On fifty-one plates, royal quarto, 1l. 1s.

Grundon's *Convenient and Ornamental Architecture*; consisting of original designs for plans, elevations, and sections, beginning with the Farmhouse, and regularly ascending to the most grand and magnificent Villa; calculated both for town and country, and to suit all persons in every station of life; with a reference and explanation in letter-press, of the use of every room in each separate building, and the dimensions accurately figured on the plans, with exact scale for the measurement; elegantly engraved on seventy copper-plates, bound, 16s.

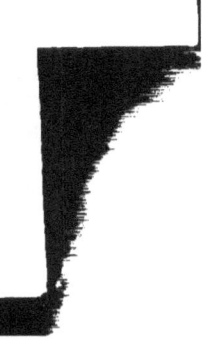

2,

bſcribers.

nber be paid

ιe Subscribers

ress will ex-

Remainder of

ιext Number,

d the History

together with

"s.

A Series of Plans for Cottages or Habitations of the Labourer, either in Husbandry, or the Mechanic Arts, adapted as well to Towns, as to the Country. To which is added, an Introduction, containing many useful Observations on the Class of Dwelling, tending to the Comfort of the Poor and Advantage of the Builder—with Calculations of Expences. By the late Mr. J. Wood, of Bath, Architect. A new Edition, with 30 Plates, large quarto, 15s. in boards.

The Country Gentleman's Architect, in a great variety of new designs for cottages, farm-houses, country-houses, villas, lodges for park or garden entrances, and ornamental wooden gates; with plans of the offices belonging to each design; distributed with a strict attention to conveniency, elegance, and economy. Engraved on thirty-two quarto plates, from designs drawn by J. Miller, Architect, sewed, 10s. 6d.

Vitruvius Britannicus, 3 vols. half bound, 6l. 6s.

The Continuation to Ditto, 2 vols.

Chambers's (Sir William) Treatise on the Decorative Parts of Civil Architecture, 3d edit. half bound, 3l. 13s. 6d.
———— Buildings and Views of Kew Gardens, half bound, 2l. 10s.
———— Designs for Chinese Buildings, &c.

Inigo Jones's Designs, by Kent, 2 vols. folio.

Leoni's Designs in Architecture, half bound.

Paine's Plans, Elevations, &c. of Noblemen's Seats, &c. folio, 2 vols. half-bound, 6l. 16s. 6d.

Plans and Elevations of Holkham-Hall in Norfolk, on 5 plates, 2l.

Ruins of Athens, by Stuart, 3 vols. of Balbec, Palmyra, Paestum, Ionia, de la Grece, par Le Roy, &c.&c.

Richardson on the Five Orders, folio, boards, 1l. 11s. 6d.
———— Plans, Elevations, &c. folio, 3l. 13s. 6d.

Nesbet's Plans for Houses, octavo, boards, 5s.

Newton's Translation of Vitruvius, 2 vols. folio.

A Treatise on Theatres, including some Experiments on Sound by G. Saunders, with plates, quarto, boards, 10s. 6d.

Perrout for the Poor, 2 tom.

Belidor l' Architecture Hydraulique, 4 tom. quarto.

Nouvelle Arch. Hydraulique, par de Prony, tome premier, quarto.

Piranesi Works, compleat, 20 vols. large folio.

Dictionare D'Architecture, Civile, Militaire et Navale, par Roland, 3 tom. quarto, with 100 Plates, 2l. 12s. 6d.

Dr. Brook Taylor's Method of Perspective made easy, both in Theory and Practice: in two Books; being an attempt to make the Art of Perspective easy and familiar, to adapt it entirely to the Arts of Design, and to make it an entertaining Study to any Gentleman who shall chuse so polite an amusement. By Joshua Kirby. Illustrated with thirty-five copper-plates, correctly engraved under the Author's inspection. The third edition, with several Additions and Improvements. Elegantly printed on imperial paper, half bound, 1l. 10s.

The same Work in two Volumes quarto, sewed, 1l. 1s.

The Perspective of Architecture, a work entirely new: deduced from the principles of Dr. Brook Taylor, and performed by two rules of universal application: Illustrated with seventy-three plates. Begun by command of his present Majesty when Prince of Wales. By Joshua Kirby. Elegantly printed on imperial paper, 1l. 16s. half bound.

The Description and Use of a new Instrument called the Architectonic Sector, by which any part of architecture may be drawn with facility and exactness. By Joshua Kirby. Illustrated with twenty-five plates. Elegantly printed on imperial paper, half bound, 1l. 1s.

Kirby's Works on Perspective, 3 vol. complete, uniformly half-bound, 4l. 4s.

The two Frontispieces, by Hogarth, to Kirby's Perspective, may be had separate, each 2s.

The Student's Instructor, in drawing and working the five Orders of Architecture; fully explaining the best methods for striking regular and quirked Mouldings, for diminishing and gluing of Columns and Capitals for making the true Diameter of an Order at any given height, for finking the Ionic Volute circular and elliptical, with finished examples on a large scale of the Orders, their Planceers, &c. and some Designs for Door Cases, by Peter Nicholson, engraved on 23 plates octavo. Price 6s. bound.

The Carpenter's New Guide, being a complete Book of Lines for Carpentry and Joinery, treating fully on Practical Geometry, Soffits, Lines for Roofs and Domes, with a great variety of Designs for Roofs, Trusse Centres, Floors, Domes, Bridges, &c. Stair-Cases and Hand-Rails of various constructions, Angle-Bars for Shop Fronts, and Raking Mouldings, with many other things entirely new. The whole founded on true Geometrical Principles; The Theory and Practice well explained and fully exemplified on 78 Copper-Plates, including some Practical Observations and Calculations on the Strength of Timber, by Peter Nicholson, bound, 12s.

The Carpenter's and Joiner's Repository; or, a new System of Lines and Proportions for Doors, Windows, Chimneys, Cornices, and Mouldings, for finishing of Rooms, &c. &c. A great variety of Stair Cases, on a plan entirely new, and easy to be understood. Circular circular Soffits, Bowing and winding in height and circular Walls, Groins, Angle Brackets, circular and elliptical Sky-Lights, and the method of Squaring and preparing their circular Bars, Shop Fronts, &c. By W. Pain, Joiner. Engraved on fifty-nine folio copper-plates, bound, 18s.

Pain's British Palladio, or the Builder's General Assistant; demonstrating, in the most easy and practical method, all the principal rules of Architecture, from the ground plan to the ornamental finish. Illustrated with several new and useful designs of houses, with their plans, elevations, and sections. Also clear and ample instructions annexed to each subject in letter-press; with a list of prices for materials and labour, and labour only. This work will be universally useful to all carpenters, bricklayers, masons, joiners, plaisterers, and others, concerned in the several branches of building, &c. comprehending the following subjects, viz. Plans, elevations, and sections of gentlemen's houses. Designs for doors, chimneys, and ceilings, with their proper embellishments, in the most modern taste. A great variety of mouldings, for tafs and furbase architraves, imposts, friezes, and cornices, with their proper ornaments for practice, drawn to half floats to which are added, scales for enlarging or lessening at pleasure, if required. Also, great variety of stair-cases; shewing the practical method of executing them, in any craft required, viz. groins, single-brackets, circular circular flowing and winding soffits, domes, sky-lights, &c. all made plain and easy on the newest capacity. The proportion of windows for the lights to rooms. Preparing foundations; the proportion of chimneys to rooms, and sections of flews. The principal timbers properly laid out to each plan, the manner of framing the roofs, and finding the length and backing of hips, either square or bevel. Scantlings of the timbers figured in proportion to their bearing. The method of trussing girders, scarfing plates, &c. and many other articles, particularly useful to all persons in the building profession. The whole correctly engraved on forty-two folio plates, from the original designs of William and James Pain, bound, 18s.

The Practical House Carpenter; or, Youth's Instructor: containing a great Variety of useful Designs in Carpentry and Architecture; as Centering for Groins, Niches, &c. Examples for Roofs, Sky-Lights, &c. The Five Orders laid down by a New Scale of Mouldings, &c. at large, with their Enrichments. Plans, Elevations and Sections of Houses for Town and Country Lodges, Hot-Houses, Green-Houses, Stables, &c. Designs for a Church, with Plan, Elevation, and two Sections; a Altar-Piece, and Pulpit. Designs for Chimney-Pieces, Shop-Fronts, Door-Cafes. Section of a Dining-Room and Library. Variety of Stair-Cases, with many other important Articles, and useful Embellishments. To which is added, a List of Prices for Materials and Labour, Labour only, and Dry Prices. The whole Illustrated, and made perfectly easy, by 148 Copper-Plates, with Explanations to each. By William Pain, Author of a Practical Builder, and British Palladio. The Fourth Edition, with large Additions. Price 13s. bound.

N. B. This is PAIN's last work.

[3]

The Practical Builder, or Workman's General Assistant; shewing the most approved and easy methods for drawing and working the whole or separate parts of any building; as, the use of the immutable groins, angle brackets, niches, &c. semicircular arches on framing jambs, the preparing and making their soffits; rules of carpentry, to find the length and backing of straight or curved hips, trusses for roofs, domes, &c. Trussing of girders, fictions of floors, &c. The proportion of the five orders in their general and particular parts; gluing of columns; hair-cases, with their ramp and twisted rails, fixing their carriages, newels, &c. Frontispieces, chimney-pieces, ceiling, cornices, architraves, &c. in the newest taste; with plans and elevations of gentlemen's and farm houses, barns, &c. By W. Pain, Architect and Joiner. Engraved on eighty-three quarto plates, bound, 11s.——A new edition, with improvements by the Author.

SCENERY AND ANTIQUITIES
OF SCOTLAND.

This Day was published,
In Two Vols. Quarto,
Enriched with ONE HUNDRED ELEGANT PLATES of
Views, Antiquities, and Natural History,
WITH
DESCRIPTIVE AND HISTORICAL ACCOUNTS.
Price 3l. 3s. Boards: and with the PLATES of NATURAL HISTORY, coloured from the ORIGINAL DRAWINGS, 6l. 6s. Boards.

REMARKABLE RUINS,
AND
ROMANTIC PROSPECTS
OF
NORTH BRITAIN;
WITH
Ancient Monuments, and Singular Subjects
OF
NATURAL HISTORY.

By the Rev. CHARLES CORDINER, of Banff.
The ENGRAVINGS by MAZELL.

Published by I. and J. TAYLOR,
No. 56, HIGH HOLBORN, LONDON.

Where may be had, in Quarto, by the same Author, a few remaining Copies of
ANTIQUITIES AND SCENERY OF THE
NORTH OF SCOTLAND,
In a Series of Letters to Thomas Pennant, Esq. with 21 Plates.
Price 13s. in Boards.

similes of all figures and dimensions, with their flat roofs, spires, and domes; trussing girders, cutting stone ceilings, groins, &c. with their sections of a barn. Engraved on twenty-four

necessary to be understood by Builders and members and freemen. 1. Geometry, lineal, plain trigonometry. 2. Surveying of land. &c. hundred examples of floors, superficies, solids, block cornices, rustic quoins, frontispieces, *Andrea Palladio*; and by equal parts. Likeline, twisted rails, compartments, obelisks, vases, &c. by the rules of levers, pulleys, axes in peritrees, wherein the properties and posture of the 8 copper-plates, by *Batty Langley*. The fourth

baths, cabinets, pavilions, garden feats, obelisks, &c. By *John Soane*. Engraved on thirty-eight

d sections, for huts, summer and winter hermitages, rustic seats, barns, mosques, monsque pavilions with flints, irregular stones, rude branches such. By *W. Wright*, Architect. Octavo,

'tages, Cottages, &c. engraved on 25 Plates,

h, Roman, and Gothic taste. By *C. T. Over tor.*

gates, doors, rails, and bridges, in the Gothic id some specimens of rails in the Chinese taste. Sixteen plates, octavo, sewed, 2s. 6d.

of columns, doors, windows, chimney-pieces, elevations and profiles, geometrically executed on Gothic Architecture. On the plates, quarto,

for chimney-pieces and door-cases, with their busts, sub-bases, and cornices for rooms, &c. fixtures, to rooms of any size. By *N. Wallis*,

e, and Enrichments to each Design. Engraved

rural and park phaetons, whiskies, single horse

a the common causes of their breaking, and the able to prevent chimneys to the size of the 'ng, Builder, sewed, 2s. 6d.

Fuel and Stoves, illustrated with proper Figures,

struction of arches made with bricks and plaster, with some Letters that passed between the Count ave, sewed, 3s.

of Architecture are carefully explained and correctly by *William Pain*. Octavo, sewed 6s.

Grecian, Roman, and Gothic orders of Architecture on 184 copper-plates, 12mo. bound, 4s.

2,

bscribers.

nber be paid

ie Subscribers
bress will ex-
Remainder of
next Number,
d the History
together with
's.

Haskey's Complete Measurer. 3s.
Hoppus's Measurer. Tables ready cast, 2s. 6d.
Plate Glass Book, 4s.

Every Man a Complete Builder; or easy Rules and Proportions for drawing and working the several parts of Architecture. In which are given a plan, elevation, and section of the curious trussed carpenter's work erected to support the centre arch of Black-Friars Bridge, from an exact measurement. Compiled by *Edward Oakley*. Octavo, sewed, 4s. 6d.

The Joiner and Cabinet-Maker's Darling; containing sixty different designs for all sorts of frets, friezes, &c. sewed, 3s.

The Carpenter's Companion; containing thirty-three designs for all sorts of Chinese railing and gates. Octavo, sewed, 2s.

The Carpenter's Complete Guide to the whole System of Gothic Railing; containing thirty-two new designs, with scales to each. Octavo, sewed, 2s.

The Carpenter's and Joiner's Vade Mecum. By *Robert Clavering* and Company, sewed 3s.

A Geometrical Plan of the Five Orders of Columns in Architecture, adjusted by fifteen parts; whereby the utmost capacity, by inspiration, may delineate and work an entire order, or any part, of any magnitude required. On a large sheet, 1s.

[2]

A Series of Plans for Cottages or Habitations of the Labourer, either in Husbandry, or the Mechanic Arts, adapted as well to Towns, as to the Country. To which is added, an Introduction, containing many useful Observations on this Class of Building, tending to the Comfort of the Poor and Advantage of the Builder—with Calculations of Expences. By the late Mr. J. Wood, of Bath, Architect. A new Edition, with 30 Plates, large quarto, 15s. in boards.

The Country Gentleman's Architect, is a great variety of new designs for cottages, farm-houses, country-houses, villas, lodges for park or garden entrances, and ornamental wooden gates; with plans of the offices belonging to each design, distributed with a strict attention to convenience, elegance, and economy. Engraved on thirty-two quarto plates, from designs drawn by J. Miller, Architect, bound, 10s. 6d.

Vitruvius Britannicus, 3 vols. half bound, 8l. 8s.
The Continuation to Ditto, 2 vols.
Chambers's (Sir William) Treatise on the Decora
——————— Buildings and Views of
——————— Designs for Chinese Bu
Inigo Jones's Designs, by Kent, 2 vols. folio.
Lewis's Designs in Architecture, half bound.
Paine's Plans, Elevations, &c. of Noblemen's S
Plans and Elevations of Holkham-Hall in Norfolk,
Ruins of Athens, by Stuart, 3 vols. of Balbec,
Richardson on the Five Orders, folio, boards, 2l.
——— Plans, Elevations, &c. folio, 3l. 13s.
Nuthall's Plans for Houses, octavo, boards, 5s.
Newton's Translation of Vitruvius, 2 vols. folio.
A Treatise on Theatres, including some Experiments
Parrant for les Ponts, 2 tom.
Belidor l'Architecture Hydraulique, 4 tom. quart
Nouvelle Arch. Hydraulique, par de Prony, tome p
Piranesi's Works, complete, 20 vols. large folio.
Di Dionaro D'Architettura, Civile, Militaire et Na
Dr. Brook Taylor's Method of Perspective made ea
Art of Perspective easy and familiar, to such
Gentlemen who shall chuse to polite un
neatly engraved under the Author's inspect
printed on imperial paper, half bound, il
The same Work in two Volumes quarto, seve
The Perspective of Architecture, a work entirely
rules of universal application: Illustrated
Prince of Wales. By Joshua Kirby. Eleg
The Description and Use of a new Instrument or
with facility and exactness. By Joshua
half bound, ib. 1s.
Kirby's Works on Perspective, 3 vol. complete
The two Frontispieces, by Hogarth, to Kirby
The Student's Instructor, in drawing and wor
striking regular and quirked Mouldings, f
of an Order to any given height, for Arch
of the Orders, their Planceers, &c. and
Price 6s. bound.
The Carpenter's New Guide, being a comp
Geometry, Soffits, Lines for Roofs and D
Bridges, &c. Stair-Cases and Hand-Rails
with many other things entirely new,
well explained and fully exemplified on
Strength of Timber, by Peter Nicholson, h
The Carpenter's and Joiner's Repository; or, a v
and Mouldings, for finishing of Rooms,
understood. Circular circular Soffits, St
and elliptical Sky-Lights, and the met
Pain, joiner. Engraved on Forty-nine F
Pain's British Palladio, or the Builder's General
rules of Architecture, from the ground
hooks, with their plans, elevations, and f
with a list of prices for materials and labo
masons, joiners, plaisterers, and others, an
ura. Plans, elevations, and sections of pu
embellishments, in the most modern taste
and cornices, with their proper ornament
lessening at pleasure, if required. Also, g——
cast required, viz. groins, single-brackets, circular circular flowing and winding Soffits, domes, sky-lights, &c. in near plain and easy to the meanest capacity. The proportion of windows for the light to rooms. Preparing foundations; the proportion of chimneys to rooms, and sections of flows. The principal timbers properly laid out on each plan, viz. the manner of framing the roofs, and finding the length and backing of hips, either square or bevel. Scantlings of the timbers, figured in proportion to their bearing. The method of trussing girders, framing plates, &c. and many other articles, particularly useful to all persons in the building profession. The whole correctly engraved on forty-one folio copper-plates, from the original designs of William and James Pain, bound, 16s.

The Practical House Carpenter; or, Tools's Instructor; containing a great Variety of useful Designs in Carpentry and Architecture; as Centring for Groins, Niches, &c. Examples for Roofs, Sky-Lights, &c. The Five Orders laid down by a New Scale Mouldings, &c. at large, with their Enrichments. Plans, Elevations and Sections of Houses for Town and Country, Lodges, Hot-Houses, Green-Houses, Stables, &c. Design for a Church, with Plan, Elevation, and two Sections; an Altar-Piece, and Pulpit. Designs for Chimney-Pieces, shop-Fronts, Door-Cases. Section of a Dining Room and Library. Variety of Stair-Cases, with many other important Articles, and useful Embellishments. To which is added, a List of Prices for Materials and Labour, Labour only, and Day Prices. The whole illustrated, and made perfectly easy, by 148 Copper-Plates, with Explanations to each. By William Pain, Author of the Practical Builder, and British Pal a o. The Fourth Edition, with large Additions. Price 15s. bound.

N. B. This is PAIN's last work.

[3]

The Practical Builder, or Workman's General Assistant; shewing the most approved and easy methods for drawing and working the whole or separate part of any building; as, the use of the travel for groins, angle brackets, niches, &c. semicircular arches on flewing jambs, the preparing and making their foffits; rules of carpentry, to find the length and backing of ſtraight or curved hips, truſſes for roofs, domes, &c. Truſſing of girders, fiction of floors, &c. The proportion of the five orders in their general and particular parts; gluing of columns; ſtair-caſes, with their ramp and twiſted rails, fixing their carriages, newels, &c. Frontiſpieces, chimney-pieces, ceilings, cornices, architraves, &c. in the neweſt taſte; with plans and elevations of gentlemen's and farm houſes, barns, &c. By W. Pain, Architect and Joiner. Engraved on eighty-three quarto plates, bound, 12s. —— A new edition, with improvements by the Author.

The Carpenter's Pocket Directory; containing the beſt methods of framing timbers of all figures and dimenſions, with their ——— on floors, roofs in lodgements, their length and backings; truſſed roofs, ſoffits, and domes; truſſing girders, ——— for arches, vaults, &c. cutting flone ceilings, groins, &c. with their and ſections of a barn. Engraved on twenty-four 1, 4s.

ly neceſſary to be underſtood by Builders and to numbers and fractions. 2. Geometry, lineal, Plain trigonometry. 6. Surveying of land, &c. a hundred examples of Baſes, Superficies, Solids, fts, block cornices, ruſtic quoins, frontiſpieces, to Andrea Palladio; and by equal parts. Like-ciſm, twiſted rails, compartments, obeliſks, viſta alies by the force of levers, pulleys, axes in pert- fice, wherein the properties and preſſure of the 20 copper-plates, by Batty Langley. The fourth

baths, cuſtards, pavilions, garden feats, obeliſks, &c. By John Soane. Engraved on thirty-eight

al factions, for huts, ſummer and winter hermit- s, ruſtic feats, barns, mosques, mosque pavi- ted with ſeats, irregular ſtones, rude branches ſ each. By W. Wright, Architect. Octavo,

ttages, Cottages, &c. engraved on 25 Plates,

ek, Roman, and Gothic taſte. By C. T. Overton.

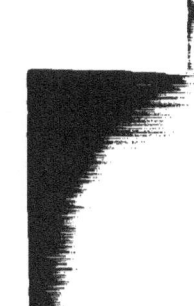

gates, doors, rails, and bridges, in the Gothic d ſome ſpecimens of rails in the Chineſe taſte, Sixteen plates, octavo, ſewed, 2s 6d.

of columns, doors, windows, chimney-pieces, nr, elevations and pieces, geometrically execu- on Gothic Architecture. On 64 plates, quarto,

, for chimney-pieces and door-caſes, with their , baſes, ſub-baſes and cornices for rooms, &c. tures, to rooms of any ſize. By N. Wallis,

s, and Enrichments to each Deſign. Engraved

road and park phaetons, whiſkies, ſingle horſe

o the common cauſes of their ſmoking, and the able to proportion chimneys to the ſize of the ing. Builder, ſewed, 2s 6d. Fuel and Stoves, illuſtrated with proper Figures,

ſtruction of arches made with bricks and plaſks, with ſome Letters that paſſed between the Count 2avo, ſewed, 2s. t of Architecture are carefully explained and exe- , by William Pain. Octavo, bound 6s. Grecian, Roman, and Gothic orders of Architec- d on 184 copper-plates, 12mo. bound, 4s.

Hoppus's Complete Meaſurer, 3s.
Hoppus's Meaſurer. Tables ready caſt, 2s. 6d.
Plate Glaſs Book, 4s.
Every Man a Complete Builder, or eaſy Rules and Proportions for drawing and working the ſeveral parts of Architecture. In which are given a plan, elevation, and ſection of the curious truſſed carpenter's work erected to ſupport the centre arch of Black-Friars Bridge, from an exact meaſurement. Compiled by Edward Oakley. Octavo, ſewed, 2s. 6d.
The Joiner and Cabinet Maker's Darling; containing ſixty different deſigns for all ſorts of freſs, friezes, &c. ſewed, 3s.
The Carpenter's Companion; containing thirty-three deſigns for all ſorts of Chineſe railing and gates. Octavo, ſewed, 2s.
The Carpenter's Complete Guide to the whole Syſtem of Gothic Railings; containing thirty-two new deſigns, with ſcales to each. Octavo, ſewed, 2s.
The Carpenter's and Joiner's Vade Mecum. By Robert Clavering and Company, ſewed 2s.
A Geometrical View of the Five Orders of Columns in Architecture, adjuſted by aliquot parts; whereby the means & capacity, by in-ſpection, may delineate and work an entire order, or any part, of any magnitude required. On a large ſheet, 1s.

[2]

A Series of Plans for Cottages or Habitations of the Labourer, either in Husbandry, or the Mechanic Arts, adapted as well to Towns, as to the Country. To which is added, an Introduction, containing many useful Observations on this Class of Building, tending to the Comfort of the Poor and Advantage of the Builder—with Calculations of Expences. By the late Mr. J. Wood, of Bath, Architect. A new Edition, with 30 Plates, large quarto, 15s. in boards.

The Country Gentleman's Architect, in a great variety of new designs for cottages, farm-houses, country-houses, villas, lodges for park or garden entrances, and ornamental wooden gates; with plans of the offices belonging to each design, distributed with a strict attention to convenience, elegance, and economy. Engraved on thirty-two quarto plates, from designs drawn by J. Miller, Architect, sewed, 10s. 6d.

Vitruvius Britannicus, 3 vols. half bound, 6l. 6s.
The Continuation to Ditto, 2 vols.
Chambers's (Sir William) Treatise on the Decor—
———————— Buildings and Views of
———————— Design for Chinese B
Inigo Jones's Designs, by Kent, 2 vols. folio.
Leoni's Designs in Architecture, half bound.
Paine's Plans, Elevations, &c. of Noblemen's B
Plans and Elevations of Holkham-Hall in Norfolk
Rules of Athens, by Stuart, 2 vols. of Balbec,
Richardson on the Five Orders, folio, boards, 2l.
———— *Plans, Elevations,* &c. folio, 3l. 13s.
Nesbit's Plan for Houses, octavo, boards, 5s.
Newton's Translation of Vitruvius, 2 vols. folio.
A Treatise on Theatres, including some Experiments
Pavement for the Poets, 2 tom.
Belidor l'Architecture Hydraulique, 4 tom. quart
Nouvelle Arch. Hydraulique, par de Prony, tome p
Piranesi's Works, compleat, 20 vols. large folio.
Dictionary D'Architecture, Civile, Militaire et Na
Dr. Brook Taylor's Method of Perspective made e
Art of Perspective easy and familiar, to ad
Gentlemen who shall choose to pursue it
roully engraved under the Author's inspect
printed on imperial paper, half bound, 1l
The same Work in two Volumes quarto, sewed
The Perspective of Architecture, a work entirely
rules of universal application: Illustrated
Prince of Wales. By *Joshua Kirby*. Elq
The Description and Use of a new Instrument or
with facility and exactness. By *Joshua*
half bound, 1l. 1s.
Kirby's Works on Perspective, 3 vol. complete
The two Frontispieces, by Hogarth, to Kirby
The Student's Instructor, in drawing and work
finking regular and quirked Mouldings, t
of an Order to any given height, for fitting
of the Orders, their Mancours, &c. and
Price 6s. bound.
The Carpenter's New Guide, being a comp
Geometry, Soffits, Lines for Roofs and D
Bridges, &c. Stair-Cases and Hand-Rails
with many other things entirely new,
well explained and fully exemplified on
Strength of Timber, by Peter Nicholson, b
The Carpenter's and Joiner's Repository, &c. 2 r
and Mouldings, for finishing of Rooms,
understood. Circular circular Soffits, &
and elliptical Sky-Lights, and the math
Pain, Joiner. Engraved on forty-nine f
Pain's British Palladio, or the Builder's General
rules of Architecture, from the ground
houses, with their plans, elevations, and
with a list of prices for materials and labo
trusses, joiners, plasterers, and others, to
ers. Plans, elevations, and sections of g
embellishments, in the most modern taste
and cornices, with their proper ornaments
lessening at pleasure, if required. Also, g
cafe required, viz. grates, angle-brackets, ... and winking forms, domes, sky-lights, &c. all made plain and easy to the meanest capacity. The proportion of windows for the light in rooms. Preparing foundations; the proportion of chimneys to rooms, and lessen of flues. The principal timbers properly laid out on each plan, viz. the manner of framing the roofs, and finding the length and backing of hips, either square or bevel. Scantlings of the timbers, figured in proportion to their bearing. The method of trussing girders, Girting plates, &c. and many other articles, particularly useful to all persons in the building profession. The whole correctly engraved on forty-two folio copper-plates, from the original designs of *William* and *James Pain*, bound, 18s.

The Practical House Carpenter, or, Youth's Instructor: containing a great Variety of useful Designs in Carpentry and Architecture, as Centering for Groins, Niches, &c. Examples for Roofs, Sky-Lights, &c. The Five Orders laid down by a New Scale. Mouldings, &c. at large, with their Enrichments. Plans, Elevations and Sections of Houses for Town and Country. Lodges, Hot Houses, Green-Houses, Sashes, &c. Design for a Church, with Plan, Elevation, and two Sections; an Altar-Piece, and Pulpit. Designs for Chimney-Pieces, Shop-Fronts, Door-Cases. Section of a Dining-Room and Library. Variety of Stair-Cases, with many other important Articles, and useful Embellishments. To which is added, a List of Prices for Materials and Labour, Labour only, and Day Prices. The whole illustrated, and made perfectly easy, by 148 Copper-Plates, with Explanations to each. By *William Pain*, Author of the Practical Builder, and British Palladio. The Fourth Edition, with large Additions. Price 13s. bound.

N. B. This is PAIN's last work.

[3]

The Practical Builder, or Workman's General Assistant; shewing the most approved and easy methods for drawing and working the whole or separate parts of any building; as, the use of the numerical geoter, single brackets, niches, &c. semicircular arches on flewing jambs, the preparing and making their coffins; rails of carpentry, to find the length and backing of sleights or curved hips, truffes for roofs, domes, &c. Truffing of girders, section of floors, &c. The proportions of the five orders in their general and particular parts; gluing of columns, frize-ends, with their raws and twisted rails, fixing their carriages, newels, &c. Plinth-pieces, chimney-pieces, cielings, cornices, architraves, &c. in the newell rails; with plans and elevations of gentlemen's and farm houses, barns, &c. By W. Pain, Architect and Joiner. Engraved on eighty-three quarto plates, bound, 12s.——A new edition, with improvements by the Author.

The Carpenter's Pocket Directory; containing the best methods of framing timbers of all figures and dimensions, with their several parts; as floors, roofs in judgements, their length and backings; truffed roofs, firens, and domes; truffing girders, partitions, and bridges, with abutments; centering for arches, vaults, &c. cutting floor cellings, groins, &c. with their moulds: centers for drawing Gothic arches, ellipses &c. With the plan and sections of a barn. Engraved on twenty-four plates, with explanations. By W. Pain, Architect and Carpenter, bound, 4s.

The Builder's Complete Assistant; or, a Library of Arts and Sciences, absolutely necessary to be understood by Builders and Workmen in general, viz. 1. Arithmetick, vulgar and decimals, in whole numbers and fractions. 2. Geometry, lineal, superficial, and solid. 3. Architecture, universal. 4. Mensuration. 5. Plain trigonometry. 6. Surveying of land, &c. 7. Mechanic powers. 8. Hydrostaticks. Illustrated by above thirteen hundred examples of lines, superficies, solids, mouldings, pedestals, columns, pilasters, entablatures, pediments, imposts, block cornices, rustic quoins, frontispieces, arcades, porticos, &c. proportioned by modules and minutes, according to *Andrea Palladio*; and by equal parts. Likewise great varieties of truffed roofs, timber bridges, centering, arches, groins, twisted rails, compartments, shells, &c. materials for bullion, fun-dials, floats, &c. and methods for raising heavy bodies by the force of levers, pulleys, axes in peritrochia, screws, and wedges; as also water by the common pump, crane, &c. wherein the properties and pressure of the air on water, &c. are exhibited. The whole exemplified on 77 large quarto copper-plates, by *Batty Langley*. The fourth edition, 2 vol. royal octavo, bound, 12s.

Designs in Architecture, consisting of plans, elevations, and sections for temples, baths, cassinos, pavilions, garden seats, obelisks, and other buildings; for decorating pleasure-grounds, parks, forests, &c. &c. By *John Soane*. Engraved on thirty-eight copper-plates, imperial octavo, sewed, 6s.

Grotesque Architecture, or Rural Amusements; consisting of plans, elevations, and sections, for huts, summer and winter hermitages, retreats, terminaries, Chinese, Gothic, and natural produce, cascades, rustic feats, baths, mosques, moresque pavilions, grotesque feats, green-houses, &c. many of which may be executed with flints, irregular stones, rude branches and roots of trees, containing twenty-eight new designs, with scales to each. By W. Wrighte, Architect. Octavo, sewed, 4s. 6d.

Ideas for Rustic Furniture, proper for Garden Chairs, Summer Houses, Hermitages, Cottages, &c. engraved on 25 Plates, Octavo. Price 4s.

The Temple Builder's most useful Companion; containing original designs in the Greek, Roman, and Gothic taste. By C. T. Overton. Engraved on fifty copper-plates, octavo, sewed, 7s.

The Carpenter's Treasure; a collection of designs for temples, with their plans, gates, doors, rails, and bridges, in the Gothic taste, with centers at large for striking Gothic curves and mouldings, and some specimens of taste in the Chinese manner, forming a complete system for rural decorations. By N. Wallis, Architect. Sixteen plates, octavo, sewed, 2s. 6d.

Gothic Architecture improved by Rules and Proportions in many grand designs of columns, doors, windows, chimney-pieces, arcades, colonnades, porticos, umbrellos, temples, pavilions, &c. with plans, elevations and profiles, geometrically examplified. By B. & T. Langley. To which is added, an Historical Discourse on Gothic Architecture. On 64 plates, quarto, bound, 15s.

The Modern Joiner; or, a Collection of original Designs, in the present taste, for chimney-pieces and feat-cafes, with their mouldings and enrichments at large; bases, tablets, ornaments for pilasters, bases, sub-bases and cornices for rooms, &c. with a Table, shewing the proportions of chimneys, with the establatures, in rooms of any size. By N. Wallis, Architect, quarto, 8s.

Outlines of Designs for Shop Fronts and Door Cases, with the Mouldings at large, and Enrichments to each Design. Engraved on 24 Plates, quarto, 3s.

Carrus Civilis; or General Designs for coaches, chariots, post-chaises, vis-à-vis, road and park phaetons, whiskies, single horse chaises, &c. Engraved on thirty plates, quarto, sewed, 10s. 6d.

An Essay on the Construction and Building of Chimneys, including an enquiry into the common causes of their smoking, and the most effectual remedies for removing so intolerable a nuisance; with a Table to proportion chimneys to the size of the room. Illustrated with proper figures. A new edition. By Robert Clavering, Builder, sewed, 2s. 6d.

Observations on Smoky Chimneys, their Causes and Cure, with Considerations on Fuel and Stoves; illustrated with proper Figures, by B. Franklin, LL.D. 1s. sewed.

This Work, with Clavering,'s Essays, may be had together, in boards, 4s. 6d.

The Manner of securing all sorts of Buildings from Fire; a treatise upon the construction of arches made with bricks and plaster, called flat arches; and of a roof without timber, called a brick roof: with some Letters that passed between the Count D'Espie, Peter Wyche, and William Hertsford, Esqrs. on this subject. Octavo, sewed, 2s.

The Builder's Pocket Treasure, in which not only the Theory but the Practical Part of Architecture are carefully explained and correctly engraved on fifty-five Copper-plates, with printed Explanations to each, by B'sham Pain. Octavo, bound 6s.

Langley's Builder's Directory, or Bench Mate; being a pocket treasury of the Grecian, Roman, and Gothic orders of Architecture, made easy to the meanest capacity, by near 500 examples, engraved on 184 copper plates, 12mo. bound, 4s.

Langley's Builder's Jewel, bound, 4s. 6d.

Hoppus's Complete Measurer, 7s.

Hoppus's Measurer. Tables ready cast, 2s. 6d.

Plate Glass Book, 4s.

Every Man a Compleat Builder; or easy Rules and Proportions for drawing and working the several parts of Architecture. In which are given a plan, elevation, and section of that curious truffed carpenter's work erected to support the center arch of Black-Friars Bridge, from an exact measurement. Compiled by *Edward Oakley*. Octavo, sewed, 2s. 6d.

The Joiner and Cabinet Maker's Darling; containing sixty different designs for all sorts of frets, fritzes, &c. sewed, 3s.

The Carpenter's Companion; containing thirty-three designs for all sorts of Chinese railing and gates. Octavo, sewed, 2s.

The Carpenter's Complete Guide to the whole System of Gothic Railing; containing thirty-two new designs, with scales to each. Octavo, sewed, 2s.

The Carpenter's and Joiner's Vade Mecum. By Robert Clavering and Company, sewed 2s.

A Geometrical View of the Five Orders of Columns in Architecture, adjusted by elegant parts; whereby the meanest capacity, by inspection, may delineate and work an entire order, or any part, of any magnitude required. On a large sheet, 1s.

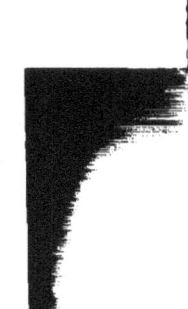

[4]

Elevation of the New Bridge at Black-Friars, with plan of the foundation and superstructure. By R. Baldwin. 12 inches by 18 inches, 5s.

Plans, Elevations and Sections of the Machines and Centering used in erecting Black-Friars Bridge; drawn and engraved by R. Baldwin, clerk of the work; on seven large plates, with explanations 10s. 6d. or with the elevation 15s.

Elevation of the Stone Bridge built over the Severn, at *Shrewsbury*; with plan of the foundation and superstructure, elegantly engraved by Roder. 1s. 6d.

A Treatise on Building in Water. By G. Semple. Quarto, with 63 plates, sewed, 12s.

Plans, Elevations, and Sections of the Gaol, Bridewell, and Sheriff's Ward, lately built at *Bodmin*, in the county of Cornwall, by *John Call*, Esq. upon the plan recommended by *John Howard*, Esq. On a large sheet, 2s. 6d.

London and Westminster improved. Illustrated by plans. To which is prefixed, a Discourse on public Magnificence; with observations on the State of Arts and Artists in this kingdom; wherein the Study of the politer arts is recommended as necessary to a liberal education: contained by some proposals relative to places not laid down in the plans. By *John Gwynn*, Architect. Boards, 5s.

Plans, Elevations, and Sections, presented to the *corporation of Bath*, for the improvement of the *Baths* in that city, intending to make the whole one grand, uniform, elegant, and convenient Structure of the Ionic order. By the late R. Dingley, Esq. Engraved on nine folio plates, by Roder, &c. sewed, 6s.

Examples, or Count Caylus's Method of painting in the manner of the Ancients. By J. H. Muntz. Octavo, bound, 5s.

The Young Draftsman's Guide to the true Outlines of the Human Figure; or a great variety of easy examples of the Human Body; calculated to encourage young beginners, and thereby lead to the habit of drawing with accuracy and facility on true principles. By an eminent Artist, deceased. Engraved on eighteen copper-plates, folio, sewed, 5s.

BOOKS of ORNAMENTS, &c.

Ornaments Displayed, on a full size for working, proper for all Carvers, Painters, &c. containing a variety of accurate Examples of Foliage and Friezes, elegantly engraved in the manner of Chalks, on 33 large folio plates, sewed, 15s.

A New Book of Ornaments; containing a variety of elegant designs for modern pannels, curiously executed in Rocco, wood, or painting, and used in decorating principal rooms. Drawn and etched by P. Columbani. Quarto, sewed, 7s. 6d.

A Variety of Capitals, Prizes, and Cornices; how to increase or decrease them, still retaining the same proportion as the original. Likewise, twelve designs for chimney-pieces, drawn 10 inch and a half to a foot. On twelve plates, drawn and etched by P. Columbani. Folio, sewed, 6s.

The Principles of Drawing Ornaments made easy, by proper examples of leaves for mouldings, capitals, scrolls, bases, foliage, &c. engraved in imitations of drawings, on fifteen plates. With instructions for learning without a master. Particularly useful to carvers, cabinet-makers, stucco-workers, painters, smiths, and every one concerned in ornamental decorations. By an Artist. Quarto, sewed, 4s. 6d.

Ornamental Iron Work, or designs in the present taste, for fan-lights, stair-case railing, window guard irons, lamp irons, palisades and gates. With a scheme for adjusting designs with facility and accuracy to any shape. Engraved on 21 plates, quarto, sewed, 6s.

A new Book of Ornaments, by S. Alba. On six plates, sewed, 2s. 6d.

Twelve new Designs of Frames for Looking-Glasses, Pictures, &c. by S. H. carver, sewed, 2s.

A Book of Tables, done in the full size commonly used for chimney-pieces. Designed and etched by J. Parker, on six plates, sewed, 3s. 6d.

Ornaments in the Palmyrene Taste, engraved on twelve quarto plates. By N. Wallis, sewed, 4s. 6d.

Law's new Book of Ornaments, sewed, 2s.

A Book of Vases, by T. Law, sewed, 2s.

A Book of Vases, by P. Columbani, sewed, 2s.

A new Book of *Eighteen Vases*, Modern and Antique, 2s.

A Book of Vases from the Antique, on twelve plates, sewed, 2s.

Gerard's new Book of Foliage, sewed, 2s.

A small Book of Ornaments, on six leaves, by G. Edwards, 1s.

A new Book of Designs for Girandoles and Glass Frames. By B. Pastorini, on ten plates, sewed, 4s.

An Interior View of Durham Cathedral, and a View of the elegant Gothic Shrine in the same. Elegantly engraved on two large Sheets, size 19 and a half by 22 and a half; the pair, 12s.

An Exterior and Interior View of St. Giles's Church in the Fields, elegantly engraved by Walker, size 18 inches by 15; the pair, 5s.

A Plan and Elevation of the King of Portugal's Palace at Mafra, on two large sheets, 6s.

A north-west View of Greenwich Church, 1s.

An elegant engraved View of the *Monument* at London, with the parts geometrically; size 21 by 33 inches; from an original by Sir C. Wren; and an Historical Account in letter press, 7s. 6d.

Sir Christopher Wren's Plan for rebuilding the City of London after the great Fire 1666, 2s.

Plan and Sections of a curious *Sailing Machine*, neatly coloured, 5s.

The Art of Practical Measuring by the Sliding Rule; shewing how to measure timber, stone, board, glass, painting, &c. also gauging, &c. By H. Coggeshall. A new edition, by J. Ham, bound, 1s.

The *Building Act* of the 14th Geo. III. With plates, shewing the proper thickness of party walls, external walls, and chimneys. A complete index, list of surveyors and their residence, &c. In a small pocket size, sewed, 2s. 6d.

N. B. The notice and certificate required by the above act, may be had printed with blank spaces for filling up, price 2d. each, or 13 for 2s.

Animals drawn from Nature, and engraved in Aquatinta, by Charles Catton, on 36 plates folio, 2l. 17s. in boards.

Smeaton's Experiments on Under-shot and Over-shot Water Wheels, &c. Octavo, with 5 Plates, boards 4s. 6d.

A General History of Inland Navigation, Foreign and Domestic; containing a complete Account of the Canals already executed in England, with Considerations on those projected. To which are added, Practical Observations. Illustrated with four Plates of Locks, Bridges, &c. and a Large Map of England coloured, shewing the Lines of the Canals executed, those projected, and the Navigable Rivers. A new Edition, with Two Addenda which complete the History to 1795. Boards, 1l. 1s.

N. B. The Addenda may be had separate, by former purchasers of the work. Also,

The Map may be had separate, price 5s. coloured.

Batalha, No. 2,

Price Fifteen Shillings to Subfcribers.

N. B. It is requefted that each Number be paid for on delivery.

ADVERTISEMENT.

THE Author begs leave to inform the Subscribers to this Work, that the whole of the Letter-press will extend to about One Hundred Pages. *The Remainder of the Introduction will be comprised in the next Number, which will be published in August next; and the History of Batalha, from* Father Luis de Sousa, *together with the Author's Notes in the subsequent Numbers.*

No. 83, Great Portland-Street,
Portland-Place, May 4, 1793.

The WORK will chiefly contain as follows:

Plans, Elevations, and Sections of the Church and Monastery.
The Manner of proportioning the Church.
The Mausoleum of its Founder, King John the First.
The Effigies of King John I. and Queen Philippa his Wife.
The Mausoleum of King Emanuel.
The Principal Windows throughout the Building, with their several Plans accurat   ed.
The Columns, Bases, and Capitals of the Church.
The Cornices, Bases, Ribs of the Vaults, Mullions, Architraves, Stiles, and other Mouldings, laid down on a large Scale, and accurately figured.
The Railing of the Church and of the Chapels externally, with the carved Orbs or Pateræ of the Vaults and other ornamental Pieces.
The Monuments and Inscriptions throughout the Building.
A View of the Entrance into the Mausoleum of King Emanuel, with the Columns and Hieroglyphics that surround it, &c. &c. with every necessary Explanation relating to each Part.

SUBSCRIBERS TO THIS WORK.

Her Most Faithful Majesty the Queen of Portugal.
His Royal Highness the Prince of Brazil.
His Grace the Duke de Lafoens, Uncle to her Most Faithful Majesty, P. R. A. Lisbon. F. R. S. Lond. &c. &c.

The Right Hon. William Conyngham, M. R. I. A. F. A. S. Lond.; the Patron and Promoter of this Work. 20 Setts.
The Right Hon. the Earl of Charlemont, P. R. I. A.
The Right Hon. the Earl of Leicester, P. A. S. and F. R. S.
Sir Joseph Banks, Bart. P. R. S. and F. A. S.
Owen Salusbury Brereton, Esq. V. P. A. S. and F. R. S.
Benjamin West, Esq. P. R. A. Historical Painter to His Majesty.
The Society of Antiquaries, London.
The Royal Academy of Arts, London.
The Trustees of the British Museum.
The Royal Irish Academy.
Trinity College, Dublin.
The Dublin Society's Library.
His Grace the Duke of Norfolk, 4 Setts.
Sir John Hort, Bart.
The Most Noble the Marquis of Lansdown.
The Lord Chancellor of Ireland.
The Countess of Moira.
The Most Noble the Marquis of Waterford.
The Right Hon. Lord Conyngham.
The Right Hon. the Earl of Moira.
The Right Reverend the Lord Bishop of Corke.
The Right Hon. Lord Orford, F. A. S.
The Viscount de Anadia.
The Chevalier de Sousa, Envoy from her Most Faithful Majesty to the Court of Sweden.
The Right Hon. the Earl of Exeter.
The Right Hon. the Earl Spencer.
The Hon. Mr. Walpole, Envoy from his Britannic Majesty to the Court of Portugal.

A LIST of all the SUBSCRIBERS will be printed at the Conclusion of the Work.

London, July, 1794.

Dedicated to the Right Hon: WILLIAM CONYNGHAM.

GOTHIC ARCHITECTURE,
ACCURATELY DELINEATED.

No. III.

PLANS, ELEVATIONS, SECTIONS, and VIEWS.

OF THE

CHURCH AND ROYAL MONASTERY OF

BATALHA,

Including the MAUSOLEA of King John I. and King Emanuel.

SITUATE IN THE

PROVINCE OF ESTRE-MADURA, IN PORTUGAL;

Measured and Drawn on the Spot, in the Year 1789.

WITH AN

HISTORICAL AND DESCRIPTIVE ACCOUNT OF THIS FAMOUS
GOTHIC STRUCTURE.

TRANSLATED FROM THE PORTUGUESE OF FR. LUIS DE SOUSA;
WITH REMARKS AND OBSERVATIONS:

To which will be prefixed,

AN INTRODUCTORY DISCOURSE ON THE PRINCIPLES OF GOTHIC ARCHITECTURE.

By JAMES MURPHY, ARCHITECT.

TERMS OF THE SUBSCRIPTION.

THIS Work will confift of Five Numbers in Imperial Folio, each containing Five Plates, or more, befides Letter-Prefs, elegantly printed on a fine Vellum Paper.

The Price to Subfcribers, Fifteen Shillings per Number, befides Half-a-Guinea to be paid at the Time of Subfcribing. Each Number to be paid for on Delivery, which will take place in the Order the Work is fubfcribed for.

A Lift of the Patronizers of the Work will be given at the Conclufion.

Subfcriptions for the Work are received by Meffrs. TAYLOR, No. 56, High Holborn, where Mr. MURPHY requefts all Orders relative to the Work to be fent.

N. B. Numbers IV and V, which complete the Work, will be publifhed in the enfuing Winter.

London, January 1, 1793.

Dedicated to the Right Hon. WILLIAM CONYNGHAM.

GOTHIC ARCHITECTURE,
ACCURATELY DELINEATED.

No. IV.

Containing Five Plates, and Six Sheets of Letter-Prefs.

PLANS, ELEVATIONS, SECTIONS, and VIEWS,

OF THE CHURCH OF

BATALHA,

IN THE

PROVINCE OF ESTRE-MADURA, IN PORTUGAL;

WITH AN

HISTORY AND DESCRIPTION

By FR. LUIS DE SOUSA;

WITH REMARKS;

To which is prefixed,

AN INTRODUCTORY DISCOURSE ON THE PRINCIPLES OF GOTHIC ARCHITECTURE.

By JAMES MURPHY, Architect.

ILLUSTRATED WITH TWENTY-SEVEN PLATES.

TERMS OF THE SUBSCRIPTION.

THIS Work will confift of Five Numbers in Imperial Folio, each containing Five Plates, or more, befides Letter-Prefs, elegantly printed on a fine Vellum Paper.

The Price to Subfcribers, Fifteen Shillings per Number, befides Half-a-Guinea to be paid at the Time of Subfcribing.

Each Number to be paid for on Delivery, which will take place in the Order the Work is fubfcribed for.

A Lift of the Patronizers of the Work will be given at the Conclufion.

Subfcriptions for the Work are received by Meffrs. Taylor, No. 56, High Holborn, where Mr. Murphy requefts all Orders relative to the Work to be fent.

N. B. The Fifth and Laft Number is in great Forwardnefs, and will be publifhed with all poffible Speed.

www.ingramcontent.com/pod-product-compliance
Lightning Source LLC
Chambersburg PA
CBHW030355170426
43202CB00010B/1378